International Conference on Control Engineering and Artificial Intelligence (CCEAI 2017)

Journal of Physics: Conference Series Volume 806

Kuala Lumpur, Malaysia
13 – 15 January 2017

Editors:

Eric Gregoire
Dejan Gjorgjevikj
Ah Chung Tsoi
Yong Yue

ISBN: 978-1-5108-3681-5
ISSN: 1742-6588

Printed from e-media with permission by:

Curran Associates, Inc.
57 Morehouse Lane
Red Hook, NY 12571

Some format issues inherent in the e-media version may also appear in this print version.

Printed by Curran Associates, Inc. (2017)

For permission requests, please contact the Institute of Physics
at the address below.

Institute of Physics
Dirac House, Temple Back
Bristol BS1 6BE UK

Phone: 44 1 17 929 7481
Fax: 44 1 17 920 0979

techtracking@iop.org

Additional copies of this publication are available from:

Curran Associates, Inc.
57 Morehouse Lane
Red Hook, NY 12571 USA
Phone: 845-758-0400
Fax: 845-758-2633
Email: curran@proceedings.com
Web: www.proceedings.com

Table of contents

Volume 806

2017 International Conference on Control Engineering and Artificial Intelligence (CCEAI 2017)

13–15 January 2017, Kuala Lumpur, Malaysia

Accepted papers received: 3 February 2017
Published online: 21 February 2017

Preface

Papers

Robotics

IOP Conf. Series: Journal of Physics: Conf. Series **806** (2017) 011001 doi:10.1088/1742-6596/806/1/011001

PREFACE

CCEAI 2017 was held in Kuala Lumpur, Malaysia during January 13-15, 2017. CCEAI 2017 was organized by Asia Pacific Institute of Science and Engineering, CCEAI 2017 is organized by Asia Pacific Institute of Science and Engineering, technical sponsored by Artois University, HKSME. The conference provides a useful and wide platform both for display the latest research and for exchange of research results and thoughts in broad Control Engineering and Artificial Intelligence. The participants of the conference were from almost every part of the world, with background of either academia or industry, even well-known enterprise. The success and prosperity of the conference is reflected high level of the papers received.

The proceedings are a compilation of the accepted papers and represent an interesting outcome of the conference. This book cover 3 chapters: Robotics; Pattern Recognition, Information Retrieval and Data Mining; Applications.

We would like to acknowledge all of those who supported CCEAI 2017. Each individual and institutional help were very important for the success of this conference. Especially we would like to thank the organizing committee for their valuable advices in the organization and helpful peer review of the papers.

We sincerely hope that CCEAI 2017 will be a forum for excellent discussions that will put forward new ideas and promote collaborative researches. We are sure that the proceedings will serve as an important research source of references and the knowledge, which will lead to not only scientific and engineering progress but also other new products and processes.

Prof. Eric Grégoire, Artois University, France
Prof. Dejan GjORGjEVIKj, Ss. Cyril and Methodius University – Skopje, MKD
Prof. Ah Chung Tsoi, University of Macau, Macau
Prof. Yong Yue, Xi'an Jiaotong-Liverpool University, China

COMMITTEES

Conference Committee Co-Chairs

Prof. Datin Dr. Sameem Binti Abdul Kareem, University of Malaya, Malaysia
Prof. Eric Grégoire, Artois University, France

Program Commitee Chairs

Prof. Dejan GjORGjEVIKj, Ss. Cyril and Methodius University – Skopje, Republic of Macedonia
Prof. Ah Chung Tsoi, University of Macau, Macau

Publication Chairs

Prof.Maxim Mozgovoy, The University of Aizu, Japan
Prof.Yong Yue, Xi'an Jiaotong-Liverpool University, China

International Technical Committee

Prof. Siti Norul Huda Sheikh Abdullah, Universiti Kebangsaan Malaysia, Malaysia
Dr Mahardhika Pratama , La Trobe Universiti, Australia
Dr. Zati Hakim Binti Azizul Hasan, University of Malaya, Malaysia
Dr. Aznul Qalid Bin Md Sabri, University of Malaya, Malaysia
Dr. Norisma Binti Idris, University of Malaya, Malaysia
Dr. Woo Chaw Seng, University of Malaya, Malaysia
Dr. Syaimak Abdul Shukor, Universiti Kebangsaan Malaysia
Dr. Chenguang (Ray) Wang, Peking University, China
Dr. Erma Rahayu Binti Mohd Faizal Abdullah, University of Malaya, Malaysia
Dr. Siti Soraya Binti Abdul Rahman,University of Malaya, Malaysia
Dr. Unaizah Hanum Binti Obaidellah, University of Malaya, Malaysia
Dr. Effirul Ikhwan Bin Ramlan , University of Malaya, Malaysia
Dr. Manik Sharma, DAV University, Jalandhar, India

IOP Conf. Series: Journal of Physics: Conf. Series **806** (2017) 011003 doi:10.1088/1742-6596/806/1/011003

Peer review statement

All papers published in this volume of *Journal of Physics: Conference Series* have been peer reviewed through processes administered by the proceedings Editors. Reviews were conducted by expert referees to the professional and scientific standards expected of a proceedings journal published by IOP Publishing.

Research the Gait Characteristics of Human Walking Based on a Robot Model and Experiment

H J He[1], D N Zhang[1], Z W Yin[2] and J H Shi[1]

[1]The 21st research institute of China electronics technology group corporation, Shanghai 200230, China

[2] Institute of Precise Forming and Knowledge Based Engineering of Shanghai Jiaotong University, Shanghai 200030, China

jevls@163.com

Abstract. In order to research the gait characteristics of human walking in different walking ways, a robot model with a single degree of freedom is put up in this paper. The system control models of the robot are established through Matlab/Simulink toolbox. The gait characteristics of straight, uphill, turning, up the stairs, down the stairs up and down are analyzed by the system control models. To verify the correctness of the theoretical analysis, an experiment was carried out. The comparison between theoretical results and experimental results shows that theoretical results are better agreement with the experimental ones. Analyze the reasons leading to amplitude error and phase error and give the improved methods. The robot model and experimental ways can provide foundation to further research the various gait characteristics of the exoskeleton robot.

1. Introduction

In recent years, there are some foreign institute carrying out studying the gait training, such as Gait Simulator in Japan [1-2], LOKOMAT in Switzerland and Hapticwalker in Germany. A 6-DOF robot integrating biological, technologies of robot, and virtual reality is designed for gait training by researchers in Makikawa laboratory in Ritsumeikan University of Japan, which could do trainings of straight, uphill, downhill, up the stairs, down the stairs and turning.

A rehabilitation robot was named as LOKOMAT and its exoskeleton is only applied to some training such as running and walking because its structure was based on treadmill, which was structured by the Swiss Federal Institute of Technology Switzerland [3].

Another kind of 6-DOF robot named as Hapticwalker for gait training (Hapticwalker) had the characteristics of exoskeleton and treadmill and was researched by Charite Hospital and Fraunhofer

IPK in Berlin in Germany. Patients can use Hapticwalker to do on-line gait analysis and the pathological data offered can evaluate the effect of rehabilitation under the help of biofeedback. Hapticwalker adopted a variety of control methods which included compliance control and position control [4-5]. Above control methods were applied to different patients. For example, if some patients were more serious illness, the compliance control will be better suitable for them to do passive exercise. But if the patients were less serious illness, position control will be suitable for them to do simple exercise such as straight, uphill, downhill, up the stairs and down the stairs. The researchers proposed that different environment should be simulated for Hapticwalker such as hard pavement and soft sand [6-7].

In addition to above research, many other colleges and institutes in Japan, Russia, the U.K., German, Korea and Singapore also started their own projects [8]. At present there have been few researches on robot combining experiment and theoretical simulation.

Therefore, a theoretical model of the gait training robot is given based on its structure in this paper. The characteristics of gaint were analyzed, and a compliance control strategy is proposed for the theoretical model. With MATLAB/Simulink toolbox, the system control models are established and the simulation is carried out in the second part. An experiment is designed and introduced in the third part. In fourth part, there is a comparison between simulation result and experimental result. Finally, there is a conclusion and outlook.

2 Simulation Model

2.1 Structure of robot

Figure 1 shows a single DOF exoskeleton platform. If a wear swings his lower limb under the help of predefined motion, the wearer's lower limb will meet closely to the single DOF exoskeleton in this platform. In this structure, the left leg of exoskeleton is designed as a rigid link pivoting, so the knee joint is a rotational one for which a DC servo motor provides power. The actuator torque τ_A is produced through DC motor about point O. Finally, the total equivalent torque is expressed as the torque τ_h about the point O. The dynamic equation of wear's lower limb of exoskeleton can be shown as,

$$(I_E + I_H)\frac{d^2\theta_e(t)}{dt^2} + (D_E + D_H)\frac{d\theta_e(t)}{dt} + (M_E g l_E + M_H g l_H)\sin\theta_e(t) + C_a sign(\frac{d\theta_e(t)}{dt}) = \tau_A(t) + \tau_H(t) \quad (1)$$

Where l, M, D and I respectively represent viscous friction coefficient, lower thigh mass, length and lower thigh inerta, viscous friction coefficient and inertial moment of the exoskeleton, C_a is friction coefficient about the knee joint of the exoskeleton robot. Figure 2 (a) explains a master-slave model of the system.

Figure 1. A prototype of lower exoskeleton with the for human enhancement motion

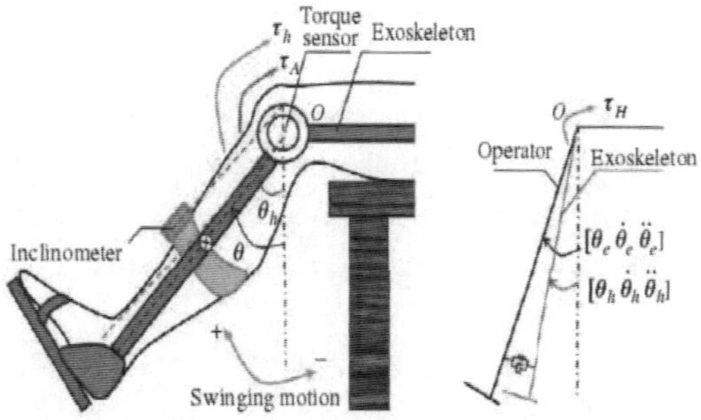

Figure 2. (a) A single DOF exoskeleton interacting wearer's lower thigh during swinging

2.2 Description of interaction resulting from human

Table 1. Nonlinear mapping models of the resulting interaction torque from human. The factors M, B, K are expressed as time varying ones

Case	Mapping model	Example
f_1	$\tau_{\text{int}} = f(\theta_h, \theta_e)$	$\tau_{\text{int}} = K(\theta_h - \theta_e)$
f_2	$\tau_{\text{int}} = f(\theta_h, \theta_e, \dot{\theta}_e, \ddot{\theta}_e)$	$\tau_{\text{int}} = M\ddot{\theta}_e + B\dot{\theta}_e + K(\theta_h - \theta_e)$
f_3	$\tau_{\text{int}} = f(\theta_h, \dot{\theta}_h, \theta_e, \dot{\theta}_e)$	$\tau_{\text{int}} = B(\dot{\theta}_h - \dot{\theta}_e) + K(\theta_h - \theta_e)$
f_4	$\tau_{\text{int}} = f(\theta_h, \dot{\theta}_h, \ddot{\theta}_h, \theta_e, \dot{\theta}_e, \ddot{\theta}_e)$	$\tau_{\text{int}} = M(\ddot{\theta}_h - \ddot{\theta}_e) + B(\dot{\theta}_h - \dot{\theta}_e) + K(\theta_h - \theta_e)$

There is an important issue that the torque τ_H is how to describe the characteristics of the interaction resulting from human. Generally, the τ_H can be described by the active torque τ_{act} from wearer's muscle torque and the torque τ_{int}. The τ_H can be expressed by a function of dynamic factors showing the human-robot structure during motion:

$$\tau_H(t)=\tau_{act}+\tau_{int} \tag{2}$$

Where, the active torque τ_{act} is dependent on the wearer central nervous system and τ_{act} is general considered as a kind of impact torque in a moment. When the impact torque is generated, the relationship between τ_H and the dynamic variables $\theta_h, \dot{\theta}_h, \ddot{\theta}_h, \theta_e, \dot{\theta}_e, \ddot{\theta}_e$ can be constructed by the interpretation of regression technique. Most of research, the environment-robot interaction is shown as many models of dynamic variables and geometric constraints. Through above models proposed in literature. The most typical model will be considered as expressed in Table 1.

2.3 Nonparametric regression techniques

Through above discussion, parameter and structure of τ_H are difficult to define explicitly. Hence, regression techniques are particularly used for this kind of regression problems. In this paper, nonparametric regression method is adopted to describe the relationship of the interaction through the dynamic variables. Gaussian Process Regression is the most standard method because of competitive performance. It has a disadvantage that Gaussian Process Regression needs a high computation cost.

Locally Weighted Projection Regression is thought as the ability of fast computation.

Gaussian Process Regression is based on Gaussian model that is explained by its covariance function and mean function in detail. It is estimated by a linear model:

$$f(x)=\psi(x)^{T}\varpi\ ,\ y= f(x)+\varepsilon \tag{3}$$

Where the weight vector is ϖ_s, the input vector is x, and additive noise is ε that is independent Gaussian distribution including variance σ_n and zero mean. With respect to linear equation (3), the regression problem is simplified as using various kernel functions such as Gaussian kernel and others. y is called as likelihood. According to the given parameters $\psi(x)$ and ϖ can be written as:

$$p\left(y|\varphi,\omega\right)\propto e^{-\frac{\left|y-\varphi^{T}\omega\right|^{2}}{2\sigma_{n}^{2}}} = N\left(\varphi^{T}\omega,\sigma_{n}^{2}I\right) \tag{4}$$

The probability density of ϖ is given by:

$$p(\omega)\propto e^{-\frac{\omega^{T}}{2}\sum_{p}^{-1}\omega} = N(0,\sum_{p}) \tag{5}$$

The probability density of the weights is given inputs and targets are proportional based on Bayes' rule:

$$p(\omega)\propto e^{-\frac{1}{2}(\omega-v)^{T}A(\omega-v)} = N(v,A^{-1}),\ A=\sigma_{n}^{-2}\varphi\varphi^{T}+\sum_{p}^{-1},\ v=\sigma_{n}^{-2}A^{-1}\varphi y \tag{6}$$

To predict the next input x_p, the f_p and V_p can be written as follow:

$$\overline{f}_{p}=k_{p}^{T}\left(K+2\sigma_{n}^{2}I\right)^{-1}y=k_{p}^{T}\xi,\quad V_{p}=k(x_{p},x_{p})-k_{p}^{T}(K,\sigma_{n}^{2}I)^{-1}k_{p} \tag{7}$$

k_p, $k(x_p,x_p)$ and k can be expressed as:

$$k_{p}=\psi_{p}^{T}\sum_{p}\varphi,\ \ k(x_{p},x_{p})=\psi_{p}^{T}\sum_{p}\psi_{p},\ K=\varphi^{T}\sum_{p}\varphi \tag{8}$$

It is known that the predicted value are respectively obtained by f_p and V_p . The vector ξ can be predicted and computed during every training.

3. Introduction of Experiment

An exoskeleton experimental platform has been designed and built for finishing the experiments with various walking ways such as straight, uphill, turning, up the stairs, down the stairs and up and down. The structure design and hardware configuration of the experimental platform are shown in Figure 3. This paper adopted the experimental platform and used the motion of the lower thigh as a scaled-down model of the swing leg in walking cycles. The influence of dynamic variables on the resulting interaction torque can be estimated.

As shown in Figure 3, the structure of the exoskeleton platform is constituted by a DC servo motor with a customized harmonic gearbox, a stationary frame and an exoskeleton arm.

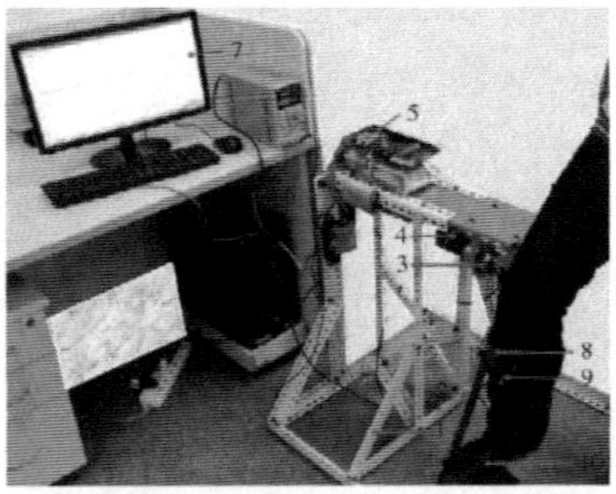

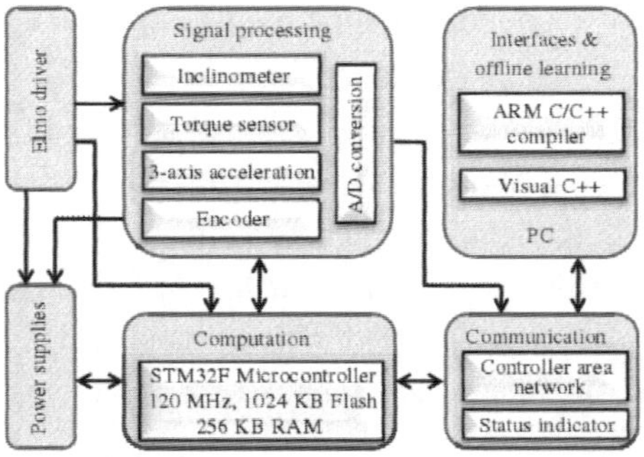

Figure 3. Experimental platform system.

Figure 4. Diagram of the distributed embedded

The motor has an incremental optical encoder of a resolution of 2000 counts per revolution that allows the measurement of the joint angle of the exoskeleton arm. In order to reduce the weight and inertia, the arm of the exoskeleton is made of aluminum. An inclinometer is built that is attached to the wearer's lower limb to measure the angular position of wearer's shank relative to gravity. A three-axis angular rate sensor which is named as L3G4200D and a three-axis accelerometer with high resolution (14 bit) is named as ADXL345 are adopted in this exoskeleton experimental platform. Besides, in order to measure the active torque from the wearer, the torque sensor is applied. It is installed between the rotating arm and the motor's shaft. To reduce the influence of error from using numerical differentiation method on the calculation of angular accelerations, digital inertial measurement units were applied to the exoskeleton and human to attenuate uncertainties occurred in the motion data.

An embedded control system is researched and designed, which increases the ability of integration, realizes the main controller to interchange information with distributed sensors favorably and ensure real-time control performance in our exoskeleton-human system laboratory. In this paper, the one DOF experimental platform is designed and its hardware configuration designed was used as a part of the whole distributed embedded system.

The hardware system is constituted by Elmo's digital servo drive for driving the DC motor, a personal computer (PC) for developing a graphical user interface and executing the online learning procedure, and a 120 MHz STM32F microcontroller as the main computational module with respect to this platform.

The message of sensor is input by the microcontroller to the microcontroller which sends appropriately computed control signals to the driver.

The communication mode is Controller Area Network CAN1 to offer a consistent and understandable behavior and visual data. The controller via CAN network (CAN2) is adopted to communicate for the Elmo drive. The general embedded system framework of the platform is shown in Figure 4.

The safe factor of the mechanical structure and the controller should be considered when the experiments are related to human motion.

4. Analysis of Theory and Experiment

4.1 Comparison between theoretical results and experimental results

According to the above the description of experiment, two persons attached with the exoskeleton perform swinging motion and the robot is controlled by a PD master-slave control. They are healthy people whose weights were 72 kg (first person) and 64.5 kg (second person). The wearers are instructed to swing their shank according to bounded sinusoidal trajectories. In order to collect data, the wearers wear the exoskeleton at the foot bracket and the inclinometer and the acceleration sensor is properly fixed at their lower thigh simultaneously. It is known that each collected data set is constituted by six inputs. From above discussion, each collected data set consists of six inputs $\phi=\left(\theta_h,\dot{\theta}_h,\ddot{\theta}_h,\theta_e,\dot{\theta}_e,\ddot{\theta}_e\right)$ and one output [τ_{int}]. A user interface is designed to store and display the collected data for every wearer on the PC.

The experiment of gait training is measured ten times for straight, uphill, turning, up the stairs, down the stairs and up and down, and the relationship can be evaluated under the same conditions for every wearer.

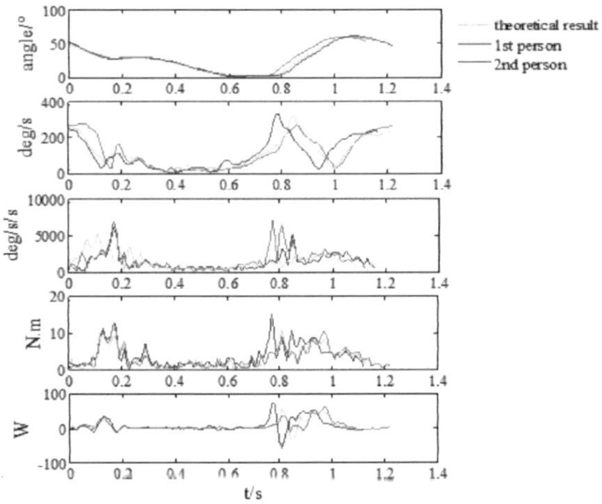

(a)

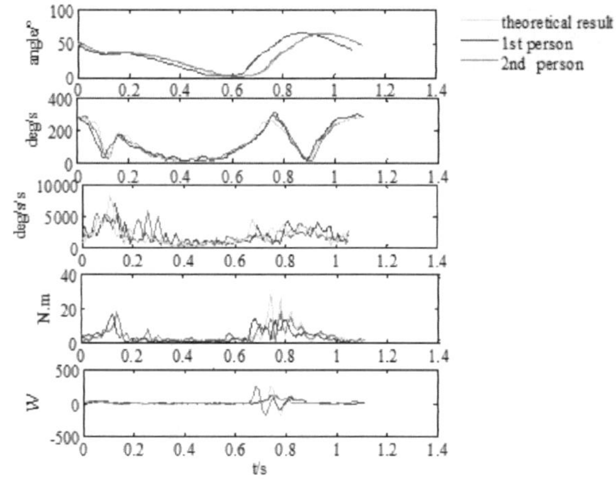

(b)

Figure 5. (a) the comparison between theoretical results and experimental results about walking along straight line;(b) the comparison between theoretical results and experimental results about Uphill

The comparison between theoretical result and experimental result of the angular position, angular velocity, angular acceleration, torque and power is respectively shown about walking along straight line in Figure 5 (a) and uphill in Figure 5 (b).

The comparison of angular velocity shows that theoretical result and experimental result have the same changing trend and the maximum amplitude error is smaller. Especially, there is no phase error.

The comparison of angular acceleration shows that theoretical result and experimental result have the same changing trend and the maximum amplitude error is smaller. The phase error is very small.

The comparison of torque shows that theoretical result and experimental result have the same changing trend and the maximum amplitude of theoretical result is larger than experimental ones.

The comparison of power shows that theoretical result and experimental result have the same changing trend and the maximum amplitude error is smaller. The phase of theoretical result is the same as the first person.

4.2 Discussion of analysis result

From the comparison between theoretical result and experimental result about different gait training such as walking along straight line, uphill, turning, up stair, down stair and up and down, changing trends of theoretical results are better agreement with experimental results. But the maximum amplitude error and phase error exists in some walking ways.

For example, the comparisons of angular position and power have a phase error and the comparisons of angular acceleration and torque have a maximum amplitude error. There are three reasons leading to amplitude error and phase error as follows:

First of all, the angular rate sensor that is used in experiment platform is not high accuracy and sensitivity. The angular rate sensor is attached with exoskeleton robot. When the wear did training with lower accuracy and sensitivity, the position feedback may be delayed. The final result will lead to phase error between theoretical result and experimental result on angular position and power.

Secondly, different wearer has different angular velocity for gait training. Because the different angular velocity produces different angular acceleration and the best torque, the vibration amplitude may be different. But the theoretical model only describe different wearer basic feature such as weight, height and basic speed, it is difficult accurate to track changing speed of wearers in real time. The final result will lead to amplitude error between theoretical result and experimental result on angular acceleration and torque.

Finally, the experimental platform may be needed to improve in order to improve the accuracy and precision of the experiment. The current experimental device did not fix vibration test sensor so that the vibration signal of gait training did not send the control system. The signal of vibration amplitude may be not evaluated accurately. The final result will lead to amplitude error between theoretical result and experimental result.

Above all, there are many factors leading to amplitude error and phase error, experimental device, wearer's training ways, sensor, control system and so on. The follow-up work will improve the theoretical model and experimental platform.

5. Conclusion and future work

In order to research the gait characteristics of straight, uphill, turning, up the stairs , down the stairs up and down, a robot model with a single degree of freedom was put up in this paper. The system control models of the robot were established through Matlab/Simulink toolbox. To verify the correctness of the theoretical analysis, an experiment was carried out. The comparison between theoretical result and experimental result about different gait training such as walking along straight line, uphill, turning, up stair, down stair and up and down, changing trends of theoretical results are better agreement with experimental results. Analyze the reasons leading to amplitude error and phase error. The follow-up work will improve the theoretical model and experimental platform. The robot model and experimental ways can provide foundation to further research the various gait characteristics of the exoskeleton robot.

References

[1] TSUJI T, TANAKA Y. Tracking control properties of human-robotic systems based on impedance control[C]//Transactions on Systems, Man and Cybernetics, Part A (Systems & Humans). Hiroshima Hiroshima University Press, 2005: 523-535.

[2] Kazerooni H, Racine J L, Huang L, et al. On the control of the Berkeley Lower Extremity Exoskeleton (BLEEX). In: Proceedings of the IEEE International Conference on Robotics and Automation, Barcelona, 2005. 4353–4360.

[3] HIDLER J M, WALL A E. Alterations in muscle activation patterns during robotic-assisted walking [J]. Clinical Biomechanics, 2005, 20(2): 184-193.

[4] TONG Xiang-qian, SHEN Ming, CHEN Gui-liang. Impedance control characteristic of active reactor [J]. Journal of Xi'an University of Technology, 2007, 23(1): 25-28.

[5] SCHMIDT H, PIORKO F, BERNHARDT R, KRUQER J, HESSE S. Synthesis of perturbations for gait rehabilitation robots [C]// Proceedings of IEEE International Conference on Rehabilitation Robotics. Chicago: IEEE Press, 2005: 74-77.

[6] PATHAK P M, MUKHERJEE A, DASGUPTA A. Impedance control of space robot[J]. International Journal of Modelling and Simulation, 2006, 26(4): 316-322.

[7] HUNT K J, JACK L P, PENNYCOTT A, PERRET C, BAUMBERGER M, KAKEBEEKE T H. Control of work rate-driven exercise facilitates cardiopulmonary training and assessment during robot-assisted gait in incomplete spinal cord injury[J]. Biomedical Signal Processing and Control, 2008, 3(1):19-28.

[8] Tran H-T, et al. Fuzzy-based impedance regulation for control of the coupled human-exoskeleton system. In: Proceedings of the IEEE Conference on Robotics and Biomimetics, Bali, 2014, in press.

Machine Vision Based Automatic Measurement Algorithm of Concentricity for Large Size Mechanical Parts

Jae Hong Shim[1] and Tae Hyeon Nam[2]

[1] Department of Mechatronics Engineering, Korea Polytechnic University, Siheung-city, Gyeonggi-do, 15073, Republic of Korea
[2] Department of Mechatronics Engineering, Graduate School, Korea Polytechnic University, Siheung-city, Gyeonggi-do, 15073, Republic of Korea

Abstract. In this paper, we proposed an algorithm of automatic measurement of concentricity for large size mechanical parts using machine vision technology. The proposed measuring method consists of the pre-processing and the measuring process. In the pre-processing process, we proposed a method for improving the image quality for accurate calculation. And in the measuring process, we proposed an efficient algorithm of concentricity measurement. Through a series of experiments, the proposed algorithm showed successfully the effectiveness of automatic measurement of concentricity for large size mechanical parts without using any additional measuring equipment.

1. Introduction

For machining, production and quality inspection of circular parts such as ring gauge, crank shaft and bore holes of automobile, retainer ring parts at semiconductor polishing process, the concentricity is very important considerations for quality control. Usually, manual dial indicator and three-dimensional measuring instrument have been used for measuring the concentricity of mechanical parts. However, if the measuring accuracy of the above instruments is μm or less, they are expensive equipment. In this paper we proposed an image-based algorithm for concentricity measurements as an inexpensive equipment [1-3].

In this paper, we obtained a concentricity of the circle by finding the center of the circle from any two points on the arc of a target circle based on the proposed algorithm proposed. In the preprocessing process, outline of the circle or the circle arc is obtained to specify two points on the arc through image processing. Through a series of experiments, the proposed algorithm showed successfully the effectiveness of automatic measurement of concentricity for large size mechanical parts without using any additional measuring equipment.

2. Pre-processing process of the input image

In this paper, we measured the concentricity of the retainer ring in CMP(Chemical-mechanical polishing) equipment for semiconductor wafer polishing as shown in figure 1.

In order to measure the concentricity of the circle of the retainer ring, it requires two points of inner circle and outer circle of the retainer ring. As the diameter of the outer circle of the retainer ring is about 360 mm, it is difficult to observe full circle within an image frame. Therefore, it should be observed in part for accurate measurement of the concentricity of large size mechanical parts as the retainer ring as shown in figure 2.

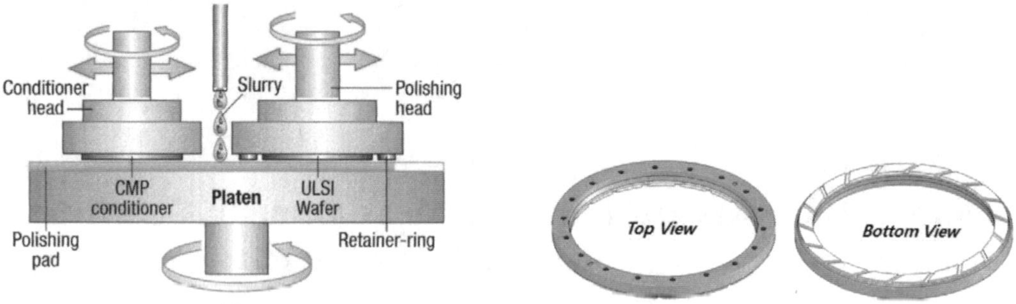

(a) (b)

Figure 1. (a) Head device of the CMP (b) retainer ring.

Figure 2. Image of section of a retainer ring.

Firstly, the object to be detected is set as a region of interest and binary operation is carried out to remove a portion other than the region of interest [4]. As the color of the objects in the image is white and the rest of image frame is black, we set the value of binary threshold as 250. When it check the binary image, clear image of circle can't be obtained because of pattern of the object surface, the influence of external interference as solar or illumination and shadows as shown in figure 3. And it is difficult to detect the correct border (edge) of the object as partly occurrence of holes.

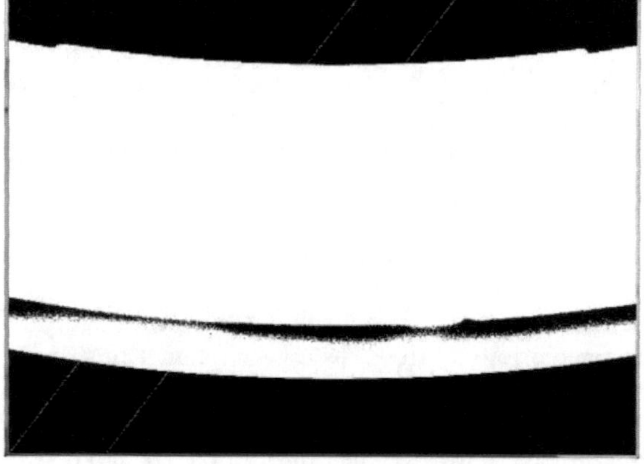

Figure 3. Firstly obtained image with noise and external influence.

The morphology technique was applied to decrease the effects of noise and holes [5-7]. As a result, the improved image was obtained as shown in figure 4.

Figure 4. Improved image by using the morphology technique.

In order to obtain any two points of the inner circle and the outer circle of the object to be measured, the edge of image corresponding to the inner and the outer circles had to be extracted. We used the canny edge algorithm for edge detection. Particularly, the retainer ring has the material characteristics that it has engraved surface pattern and so its color image is relatively evenly distributed. Canny edge detection algorithm was selected as it has better performance of edge detection than the other edge detectors and can remove all the edges associated with the gray matter of the original image. Firstly, the Sobel mask is applied as (1):

$$\nabla \mathrm{f} = \begin{bmatrix} \frac{\partial f}{\partial x} & \frac{\partial f}{\partial y} \end{bmatrix} \tag{1}$$

where *f* is an two dimensional image.

And then, non-maximum suppression process is carried out to select a local maxima value. In order to obtain any two points of the inner circle and the outer circle of the object to be measured, the edge of image corresponding to the inner and the outer circles had to be extracted. Despite the pre-processing in the image, any part with non-edge may be detected. So it needs to remove these non-maximum. As a result, clean edge of the inner and outer circles of the target object was obtained as shown in figure 5.

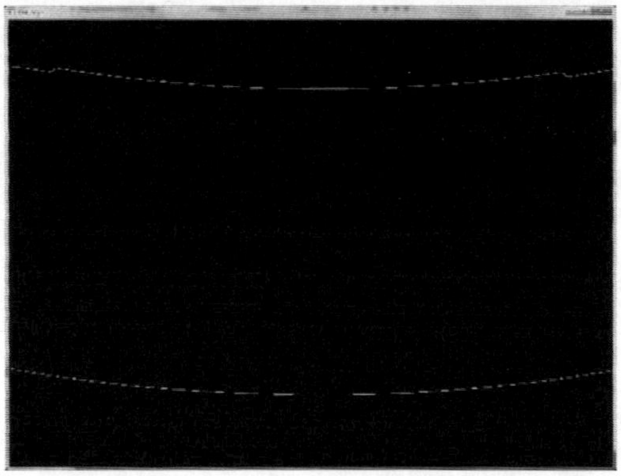

Figure 5. Edge image obtained by using the Canny edge detection algorithm.

3. Concentricity measurement algorithm

In this paper, since the diameter of a large size parts such as a retainer ring has close to 400 mm, we proposes a method of finding the center of the circle utilizing the similar triangle ratio about only a portion of the original image. Now, in order to measure the concentricity of a circular object, any two points are selected from the above edge detected image. As shown in figure 6, two y axis straight lines are drew to intervals of 200 pixels in the image and obtained the cross points of the coordinates between the y axis straight line and the edge of the circle in figure 5.

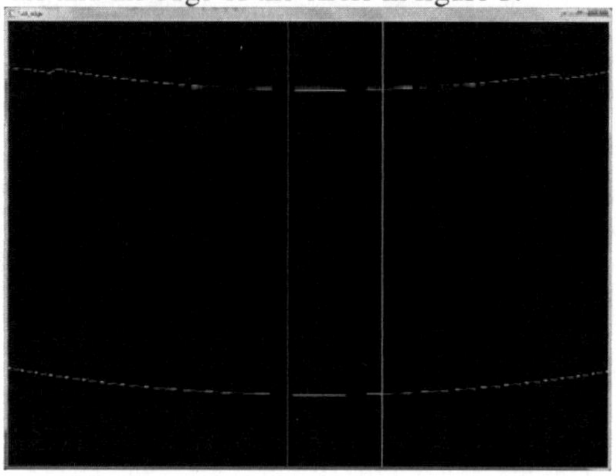

Figure 6. Cross points between two y axis straight lines and the edge detected.

Figure 7 represents the above mentioned method for finding the center of a circle.

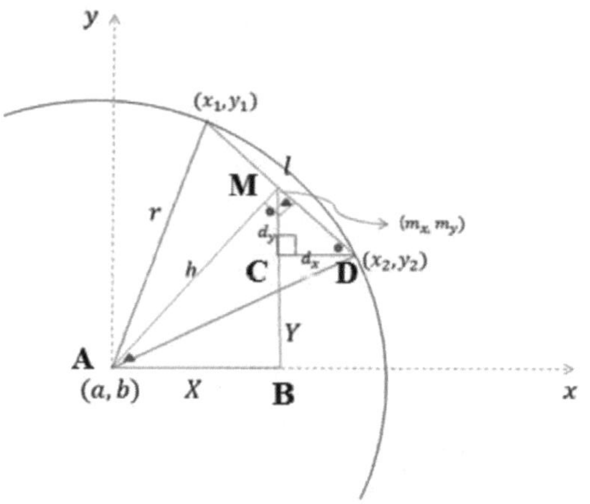

Figure 7. Schematic method of finding the center point of a circle using two points on the arc.

In order to obtain the center point (*a*, *b*), it should be find the center of the arc (m_x , m_y) and vectors *X*, *Y*. If radius of the circle is constant, △**MAB** and △**DMC** has the relation of similar triangles. Therefore, *X* and *Y* is obtained from (2) and (3), the center point *a* and *b* is found from (4) and (5):

$$X = \frac{2\, d_y\, h}{l} \tag{2}$$

$$Y = \frac{2\, d_x\, h}{l} \tag{3}$$

$$a = m_x \pm X \tag{4}$$

$$b = m_y \pm Y \tag{5}$$

4. Experimental results and discussion

In order to compare the accuracy of concentricity measurement with the proposed method, we measured concentricity of the retainer ring used in the experiment by using the high-precision contact-type three-dimensional measuring instrument of Roncorder EC 5100, and the measured concentricity was about 0.004mm. Using the proposed algorithm, the concentricity of the same retainer ring was measured as about 0.00903mm in case of using a camera having a resolution of 1280 pixels and its error range was 0.005mm. In case of using a camera having a resolution of 2560 pixels, the measured concentricity was approximately 0.00415 mm and its error range was 0.0001 mm. Table 1 shows the experimental results.

Table 1. Experimentally measured concentricity of a retainer ring by using the proposed algorithm.

Number of iteration	In case of using a camera with 1280 Pixels	In case of using a camera with 2560 Pixels
1	0.00903mm	0.004152mm
2	0.00903mm	0.004152mm
3	0.00903mm	0.007019mm
4	0.00903mm	0.004152mm
5	0.00715mm	0.004152mm
6	0.00903mm	0.005107mm
7	0.00903mm	0.004152mm
8	0.00903mm	0.004152mm
9	0.00903mm	0.004152mm
10	0.00903mm	0.004152mm

5. Conclusions

In this paper, we proposed an algorithm of concentricity measurement of large size mechanical parts by using the image sensor. Through a series of experiments, the proposed algorithm showed similar measuring accuracy compared with a high-precision contact-type three-dimensional measuring instrument(CMM). In addition, the proposed algorithm was capable of high-speed measurement within an average 0.1 seconds. Actually compared with the manual indicator or the three-dimensional measuring devices, the proposed machine vision-based concentricity measurement device is cheap and can be supported by a fast and precise measurement effect.

Acknowledgments
This paper was supported by Gyeonggi-do and Siheung-city under Industry-University Convergence Campus Outcome Utilization Project, a provincial government financial support program, Korea.

References

[1] Huh G M and Moon H Y 2003 The development of automatic measurement algorithm of concentricity and roundness using image processing technique *J. Control, Robotics and Systems* **9 3** pp 227-235

[2] Zhang T Y and Suen C Y 1984 A fast parallel algorithm for thinning digital patterns *Communications of the ACM* **27 3** pp 236-239

[3] Xie Y, Pang H, Wang C and Xaio S 2009 Small components concentricity measuring based on machine vision *Int. Symp. on Intelligent Information Systems and Applications* pp 522-525

[4] Shi B H and Wei Y H 2014 Detection method of optical fiber connector internal parts based on machine vision *Applied Mechanics and Materials* **568** pp 483-488

[5] Leta F R, Feliciano F F, Souza I L and Cataldo E 2006 Discussing accuracy in an automatic measurement system using computer vision techniques *ABCM Symposium Series in Mechatronics* **2** pp 645-652

[6] Chen M C 2002 Roundness measurements for discontinuous perimeters via machine visions *Computers in Industry* **47** pp 185-197

[7] Shankar N G and Zhong Z W 2005 Defect detection on semiconductor wafer surfaces *Microelectronic Engineering* **77** pp 337-346

IOP Conf. Series: Journal of Physics: Conf. Series **806** (2017) 012003 doi:10.1088/1742-6596/806/1/012003

A Lane Following Mobile Robot Navigation System Using Mono Camera

Yeongcheol Cho[1,a], Seungwoo Kim[2,b] and Seongkeun Park[3,c]

[1]Department of Electronic and Information Engineering, Soonchunhyang univ, Republic of KOREA
[23]Department of Smart car, Soonchunhyang univ, Republic of KOREA

[a]ksh111angel@naver.com, [b]seungwo@sch.ac.kr, [c]keiny@sch.ac.kr

Abstract. In this paper, we develop a lane following mobile robot using mono camera. By using camera, robot can recognize its left and right side lane, and maintain the center line of robot track. We use Hough Transform for detecting lane, and PID controller for control direction of mobile robot. The validity of our robot system is performed in a real world robot track environment which is built up in our laboratory.

1. Introduction

In order to provide convenience and welfare, robotics technology has been studied in many difference fields, and especially, many researchers are focused on self-driving of mobile robot navigation [1].

To achieve autonomous navigation of mobile robot, perception of surroundings, driving path planning and robot control algorithm are key features. Many robotics technology include home service robot, AGV and autonomous vehicle are based on mobile robot navigation.

Recently, as robot hardware and robotics technology have been improved, there are many researches about robot perception and recognition systems. For example, using mono camera, a kind of CCD image sensor, lane detection and obstacles detection algorithms are proposed in [2].
While lane detection and obstacles recognition using mono camera in real robot systems, it is hard to compute the distance between robot and obstacles or lane. To overcome this drawback, lidar and camera fusion system has proposed. Finally, PID controller is applied to control mobile robot

In this paper, we propose a sensor fusion system using lidar and camera and its application to robot self-driving system. Using Hough transform, we can detect the lane information in mono camera data and lidar will provide the distance between robot and obstacles. The remaining part of this paper is as follow. We explain lidar/camera recognition algorithm in chapter 2 and experimental setup and results are remarked in chapter 4. The conclusion and future works are in chapter 5.

2. Mobile Robot of Navigation System

In this chapter, we propose lane and obstacle detection system using lidar and camera. The whole system of proposed system is remarked in figure 1.

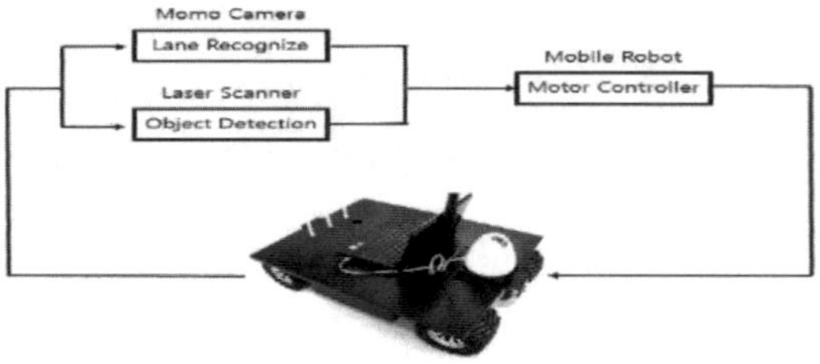

Figure 1. Robot Perception System

2.1. A Lane Recognition System of Mobile Robot

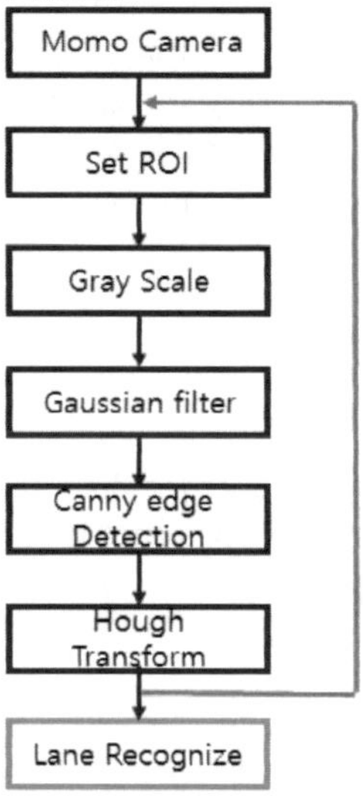

Figure 2. Lane recognition system

In this subchapter, we introduce a lane detection system using mono camera. The proposed lane detection algorithm is in figure 2. First, for reducing computation complexity, we extract region of interest(ROI) in image data and remove useless data. As converting color image to gray scale image, we accomplish more efficiency in terms of computation, and Gaussian filter is applied for remove noise in image data. And in order to find edge data, we apply canny edge detection method, and hough transform is used for lane recognition..

2.1.1 Gaussian Filter. In this paper, we applied Gaussian filter for pre-processing. Gaussian filter can remove high frequency noise in image and it can make the lane detection accuracy[3]. The 2-dimension Gaussian filter equation is as follow.

$$g(x,y) = \frac{1}{2\pi e^2} e^{\frac{x^2+y^2}{2\sigma^2}} \tag{1}$$

2.1.2 Canny Edge Detection. Canny edge detector is based on Gaussian blur and sobel masking method[4] Sobel edge masking is a kind of 1st order derivation operator, it can detect edge of every direction and normalization of every pixel data[5]. The detail of Sobel edge detector is as follow in eq. (2)

IOP Conf. Series: Journal of Physics: Conf. Series **806** (2017) 012003 doi:10.1088/1742-6596/806/1/012003

$$\nabla G(y,x)=(\frac{\partial G}{\partial},\frac{\partial G}{\partial x})=(d_x,d_y)=(G(y+1),x \quad G(y \quad 1,x),G(y,x+1) \quad G(y,x \quad 1)) \tag{2}$$

The gradient of 2-dimensional image is partial derivative of x direction and y direction.

$$S(y,x)=magnitude(\nabla G)=\sqrt{d_y{}^2+d_x{}^2} \tag{3}$$

$$D(y,x)=tan^{\ 1}(\frac{d_y}{d_x}) \tag{4}$$

The pixel of edge which is detected by sobel masking can be defined by geometric mean of longitudinal and lateral gradient, S(y,x) is magnitude of edge, and D(y,x) is direction of gradient.

2.1.3 Hough Transforms. In this paper, hough transform [6] has been applied to detection lane Random points on (x,y) co-ordinate(Cartesian coordinate) can be represented in curved line on (r,θ) coordinate (polar coordinate) and if these curved line make a cross point, Random points on (x,y) co-ordinate can be regarded as points in a same line like figure (3) So we can get one line like eq.(5) to transform (x,y) co-ordinate to (r,θ) coordinate

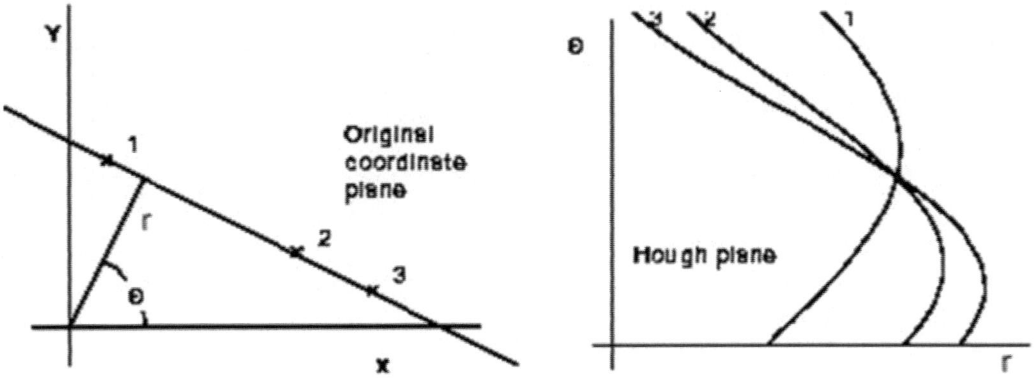

Figure 3. Hough transform

$$r=x \quad cos\theta+y \quad sin\theta \tag{5}$$

2.2. Obstacle Detection System of Mobile Robot
Lidar (a.k.a Laser scanner) transmit light to obstacle and it calculate the distance between sensor to obstacle using light velocity and time of flight. Eq.(6) is distance computing equation.

$$R=\frac{T}{2}C \tag{6}$$

R is distance to obstacles, T is time of flight, C is velocity of light

2.3. A Control System of Mobile Robot
The control system of mobile robot is as follow.

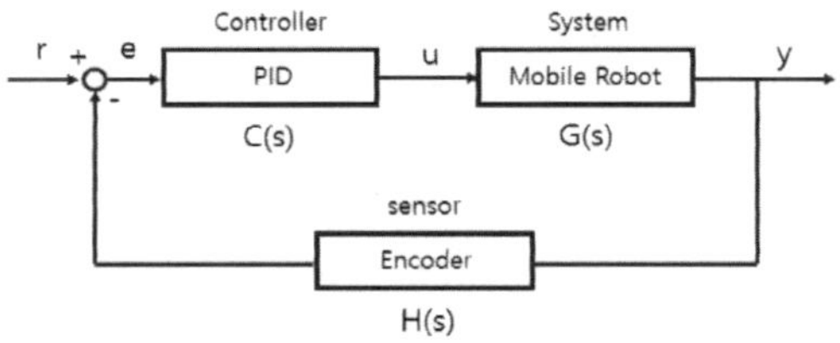

Figure 4. Control system

The robot dynamic system is as follow

$$\tau = K_t i_a = J_a \frac{d^2\theta}{dt} + B\frac{d\theta}{dt} \tag{7}$$

In eq. (7), τ is motor torque, K is torque constant, J is inertial mass of motor, B is friction coefficient of motor. θ is turning angle of motor.

$$F = \frac{T_M \quad G_r}{R_W} \tag{8}$$

In eq.(8), F is force via robot motor, G_r is gear ratio of motor and wheel, R_W is radius of robot wheel.

$$F = m\frac{dv}{dt} + bv \tag{9}$$

In eq.(9), m is robot mass, b is friction coefficient of mobile robot and V is velocity.
Finally, PID controller is represented like eq.(10).

$$u(t) = k_p e(t) + k_i \int_0^t e(\tau)d\tau + k_d \frac{de(t)}{dt} \tag{10}$$

3. Experiment and Result

3.1. Lane Recognition System

In this chapter, we show the real world experiment about our proposed method. The detail of robot in this experiment is below Figure 5 and Table 1.

Figure 5. Mobile robot platform

Table 1. Specification of the Mobile Robot

Category	Specification
Main controller	ATmega128
Communication	RS232c
Motor	2 DC motor, Servo motor
Size	650mm (length) X 470mm (width) X 158mm (height)
Weight	8Kg

Figure 6. Original image **Figure 7.** ROI image.

Figure 6 is original image of robot camera and figure 7 is ROI.

Figure 8. Gray image **Figure 9.** .Filtering image

Figure 8 is gray scale image and Gaussian filtering results is shown in figure 9. By using Gaussian filter, we can remove high frequency noise and detect lane edge more accurately.

Figure 10. Edge detection image **Figure 11.** .Hough transform image

In figure 10, the result of canny edge detector is shown and based on these edges, hough transform recognize lane. Figure 11 is hough transform result. We can verify the validity of hough transform in real world system.

Figure 12. Lane recognition

Finally, we can make reference control line using mean of right, left side lane in figure 12.

3.2. Obstacle Detection System

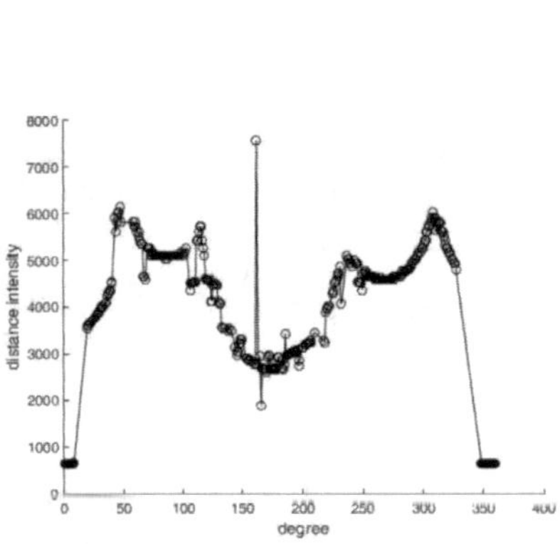

Figure 13. Laser scanner graph

Figure 13 is lidar data. As you can see in figure 13, the signal intensities in 0~10 degree and 350~360 degree are small compared to other angle. This means there are some obstacles intensities in 0~10 degree and 350~360 degree position.

4. Conclusion

In this paper, we proposed an perception system for autonomous robot navigation system using mono camera and single lidar sensor. Mono camera can detect lane information with hough transform and lidar sensor can recognize obstacles information based on TOF information. By using two sensors, we can design the robot controller for following lane track. Future works will be focused on more dynamic environement.

Acknowledgement

This work was supported by Industrial Human Resources and Skill Development Program (N0001415, Display Expert Tranning Project for Advanced Display equipments and components engineer) funded by the Ministry of Trade, Industry & Energy (MOTIE, Korea)

References

[1] Bakar, M. N. A., Nagarajan, R., & Saad, A. R. M. (2011, April). Development of a doctor following mobile robot with mono-vision based marker detection. In Applied power electronics colloquium (IAPEC), 2011 IEEE (pp. 86-91). IEEE.

[2] More Broggi, A., Cappalunga, A., Caraffi, C., Cattani, S., Ghidoni, S., Grisleri, P., & Zani, P. (2010). TerraMax Vision at the Urban Challenge 2007. IEEE Transactions on Intelligent Transportation Systems, 11(1), 194-205.

[3] Van Vliet, L. J., Young, I. T., & Verbeek, P. W. (1998, August). Recursive Gaussian derivative filters. In Pattern Recognition, 1998. Proceedings. Fourteenth International Conference on (Vol. 1, pp. 509-514). IEEE.

[4] Canny, J. (1986). A computational approach to edge detection. IEEE Transactions on pattern analysis and machine intelligence, (6), 679-698.

[5] . Jin-Yu, Z., Yan, C., & Xian-Xiang, H. (2009, April). Edge detection of images based on improved Sobel operator and genetic algorithms. In 2009 International Conference on Image Analysis and Signal Processing (pp. 31-35). IEEE.

[6] Duan, D., Xie, M., Mo, Q., Han, Z., & Wan, Y. (2010, October). An improved Hough transform for line detection. In 2010 International Conference on Computer Application and System Modeling (ICCASM 2010) (Vol. 2, pp. V2-354). IEEE.

A Vehicle License Plate Recognition System Using Morphological ROI (Region of Interest) Map Generated from Morphology Operation

Taewan Kim and Eungtae Kim

Dept. of Electronics Engineering, Korea Polytechnic University, 237 Sangidaehak-ro, Siheung-si, Gyeonggi-do, Korea

Abstract. In this paper, morphology features are used to improve license plate region and Morphological ROI (MROI) map that consists of a standard deviation map of open and close images is proposed. Unlike conventional morphology techniques, MROI map is used to improve the detection of license plate in a complex background. Furthermore, distribution of the MROI map was analyzed to determine a threshold level, generate binary code images, and finally, determine the vehicle license plate region by using the features of the plate such as aspect ratio and vertical edge intensity. The proposed method could improve the error rate by approximately 4-7% depending on the pattern of the images.

1. Introduction

The vehicle license plate recognition system is one component of the intelligent transportation system and is widely used for the tolling system on the highway and security closed-circuit television (CCTV). Recently, this system is used for autonomous vehicles.

The license plate recognition system (LPRS) consists of two main technologies, which are the license plate positioning technology and the plate number recognition technology. The former technology detects the position of the license plate in a complex background and the latter technology recognizes the plate numbers in the plate region detected from the first technology. To process the LPRS effectively, the accurate position detection method is very important.

Current plate position detecting algorithms use the edges of characters and numbers in plates[1], color feature of plates[2], sliding window technique that scans entire images of plates[3], and morphological operation [4-5].

The method using the edges of characters in plates utilize brightness vector. It uses the fact that the brightness variation of the characters in the plate is clearly distinguished than other regions. Therefore, vertical edge of the image are extracted and analyzed to detect the license plate. However, it can cause errors if there are many vertical edges in other regions than the plate region.

The method using the color feature of plates searches for pixels that are within the color range of the plate and examine them. It performs badly where there are a lot of illuminance variations.

One of the most popular methods to scan and detect number plates is the sliding windows method. It requires very complicated operations because it needs to scan the entire images.

The final method uses erosion and dilation operations through random structure element to improve the license plate region and detect license plate numbers. Most popular algorithms are the top-hat and the bottom-hat method, which are as complicated as the sliding windows method.

In this paper, we propose a new method to improve the conventional morphological methods for more accurate license plate detection in a complex background.

2. The Proposed License Plate Detection Algorithm

There are two main plate detection methods using morphological operations: top-hat method and bottom-hat method. These operations can effectively find more apparent edge regions. Because these methods are used to detect edges, they can cause errors if there are strong edges in the background.

To solve this problem, we propose the license plate detection method using Morphological ROI(MROI) map as shown in Fig.1. M-ROI map consists of the standard deviation of morphological open and close image. It uses to find to the brightness difference between the characters and the background in the plate. It is important to detect the plate regions that have high standard deviations and are clearly distinguished from the background.

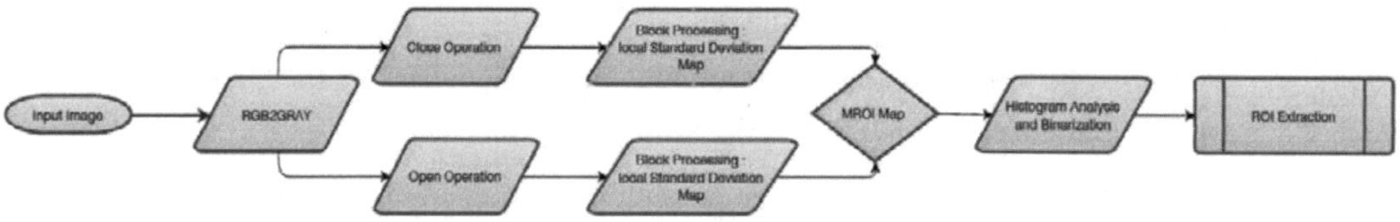

Figure 1. Flow chart of the proposed license plate detection algorithm

2.1. *Pre-Processing by Morphological Operation*

A color image is first converted to the grayscale image. The morphological operations such as the close and open operations are then performed. The size of the structure element (SE) is 4 x 4. It was determined by considering the sizes of characters and numbers in the plate. If this size is too small, the overall noise level will be larger, and if too large, the entire image will be blurred.

(a) (b) (c)

Figure 2.(a) Original Image, **(b)** Open operated Image, **(c)** Close operated Image

Fig. 2(a) shows an image that contains many strong edges. The images in Fig. 2(b) and Fig. 2(c) show the results after open and close operations, respectively. After the open operation is performed by erosion-dilation operations, black characters and numbers in the plate regions are emphasized. Moreover, through the close operation such as dilation-erosion operations, the characters and numbers in the plate regions are disappear. As shown in Fig.2, the processing results of the open and close operations are discriminated noticeably in the plate region.

2.2. *Block Processing of MROI map*

The standard deviations of the open-operated image and close-operated image are calculated based on 15x5 rectangular blocks. For the shape and size of the block, considering the blank in the center of the plate, the width was made larger than the height. As shown in Fig.3, a map is then generated by using computed standard deviations of the blocks. An element of map, W(i,j), is a standard deviation of block. If the image size is 720x480, then it is converted to 48x96 map by the block processing.

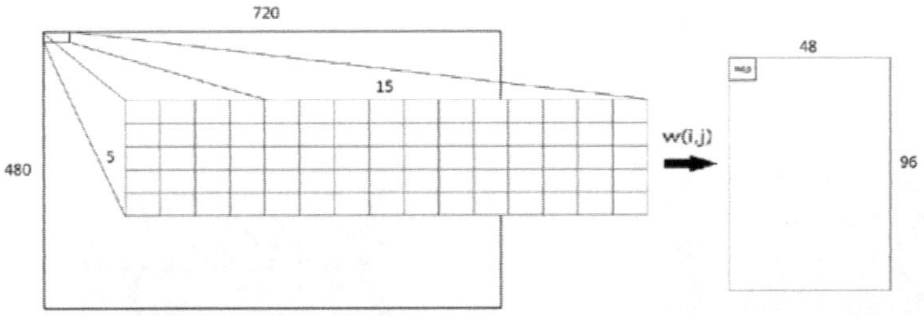

Figure 3. Generation Process of the Standard Deviation Map

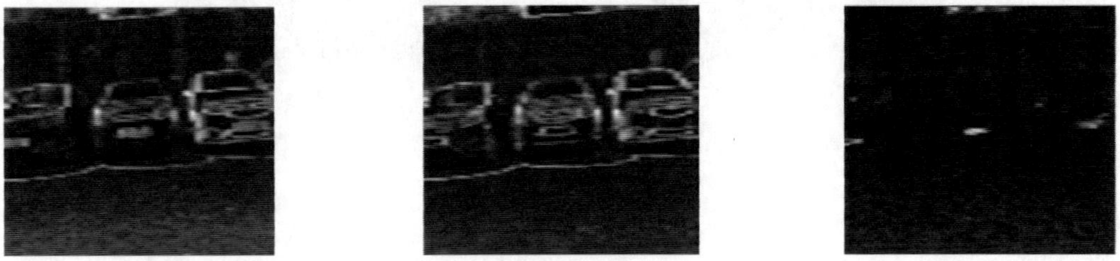

(a) **(b)** **(c)**

Figure 4.(a) Open deviation Map, **(b)** Close deviation Map, **(c)** MROI Map

The largest difference between the open deviation map and the close deviation map results can be found on the license plate. Thus, to detect the license plate region, we compute the value as in Equation (1).

$$\text{MROI Map} = \frac{(Open\ Deviation\ Map + 1)}{(Close\ Deviation\ Map + 1)} \quad (1)$$

Fig. 4(c) shows the result after Equation (1) was applied and this result has the large value of ROI map of in the plate region. The values of ROI map are very low in other regions such as backgrounds and smooth regions.

2.3. Histogram Analysis

To convert the MROI map shown in Fig. 4(c) to a binary image, a threshold value should be determined. Fig. 5 shows a histogram of the MROI map. The histogram is normalized. It shows that they are concentrated in the lower values and less concentrated in the higher values.

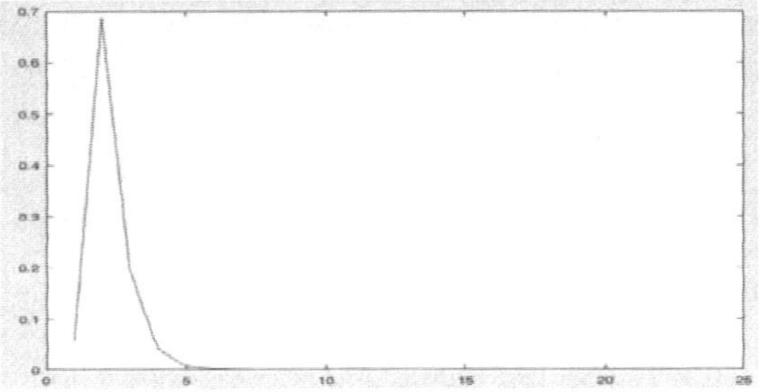

Figure 5. Typical Histogram of ROI MAP

The higher values of ROI map include high texture region such as the license plate region. In order to find the candidate of license plate, the binarization of map in performed by Thresholding. Fig. 6(a) shows the binary image of the ROI map after exclusion 99% of histogram, in which a small amount of noise is included.

IOP Conf. Series: Journal of Physics: Conf. Series **806** (2017) 012004 doi:10.1088/1742-6596/806/1/012004

To eliminate the noise, we used a simple mask. If the neighbouring standard deviation data is not connected to any of four directions, up, down, left, and right, then it is considered as a noise region.
The results are shown in Fig. 6(b).

(a) **(b)**

Figure 6.(a) Binary ROI Map, **(b)** Noise Removal Mask Image

3. ROI Verification Process

The plate region verification is processed by two methods. The first method uses the aspect ratios of the possible plate regions. The second method uses the vertical edge distribution of the possible plate regions.

3.1. Determination by Aspect Ratio

The real size of the plate is fixed. It is always either 520mm x 110mm or 335mm x 155mm. The aspect ratios for plates are 520/110=4.73, 335/155=2.16 and the range of suitable ratio is $2 \leq$ Real Aspect Ratio ≤ 5 in real image. In the case of the aspect ratio in binary ROI map, it ranges from 0.66 to 1.66, approximately 1 to 2. Then, it is available to determine whether these regions are the plate regions or not.

3.2. Determination by Vertical Edge

After the candidate region of Fig6.(b) is determined by the aspect ratio feature, those regions are selected from the original image. The characters and numbers in the real plates usually have strong vertical edges. Therefore, the vertical edges using vertical Sobel mask are calculated. Those vertical edges are projected on the vertical plane to compute the edge distribution. The vertical edges in the plate regions are concentrated in the center of the vertical plane, while the vertical edges in other regions are existed randomly. To find the higher edge value in the center, the license plate region can be determined by sum of multiplying the normalized edge histogram value with weighting coefficient value. Coefficient value is expressed as Equation (2).

4. The Experimental Result

$$coefficient\ : f(x)\begin{cases} \mathrm{x} \div \dfrac{\mathrm{row_{total}}}{2} & \left(\mathrm{if} \quad \mathrm{x} \leq \dfrac{row_{total}}{2}\right) \\ (\mathrm{row_{total}} - \mathrm{x}) \div \dfrac{\mathrm{row_{total}}}{2} & \left(\mathrm{if} \quad \mathrm{x} > \dfrac{row_{total}}{2}\right) \end{cases} \tag{2}$$

where x is row number of selected candidate region and $\mathrm{row_{total}}$ is max of row number.
In the simulation, various 720x480 real images are used as shown in Fig 7, in which some have a very simple background such as a tollgate and the other images have a relatively more complicated background such as a parking lot and multiple cars on the road.

The accuracy of the experiment is determined by implementing algorithms and indicating the plate regions with the bounding boxes. If the bounding box includes a plate region, the system determines that it worked successfully. Otherwise, the system determines that it detected a wrong region.

Figure 7. Examples of Test Data Samples

Fig.8 is the results of the proposed method in various environments. As shown in Table 1, when 30 tollgate images were tested, the plate regions were successfully detected without any errors due to the simple background. The accuracy of the proposed method is 90%, 40% in the complicated parking lot and cars on the road, respectively.

The proposed method has the better performance than the conventional method in the situation that the edges of background are complicated shown in Fig.9. The conventional method has an error to detect the vertical edges in the background and cars as well as the edges in the plate region. The proposed method detects the exact plate region even if the edges of background are strong.

Figure 8. Results Images of proposed method

Table 1. Results of the experiment

	Tollgate	Parking garage	On the Road
Number of Images	30	30	30
Conventional method Accuracy [1]	100%	86%	33%
Propose method Accuracy	100%	90%	40%

5. Conclusion

In this paper, vehicle license plate detection using Morphological ROI map was proposed in the complex vehicle images. The MROI map is made by using the standard deviation of morphological open and close images, and the threshold value is calculated using the distribution of the MROI map to effectively detect the candidate region. After detecting candidate regions, those are verified using the features of the license plate. Experimental results show that the proposed method has the higher detection rate then to conventional method by 4-7% in the complex environments.

IOP Conf. Series: Journal of Physics: Conf. Series **806** (2017) 012004 doi:10.1088/1742-6596/806/1/012004

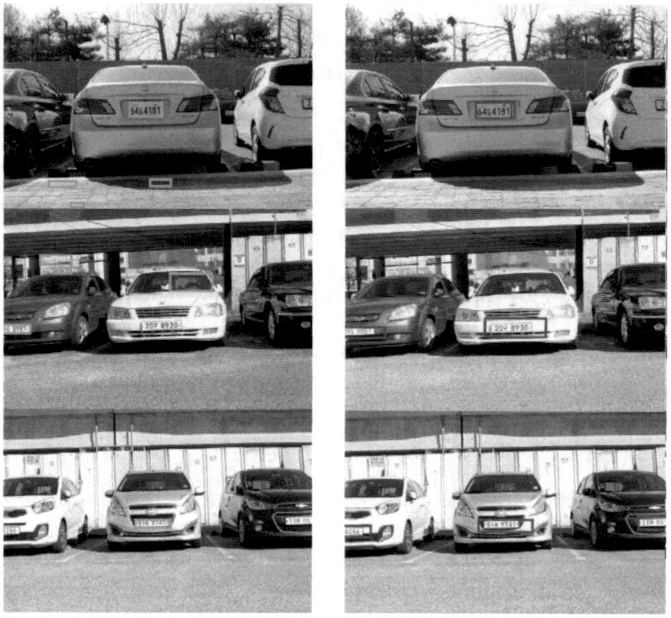

Figure 9. Comparison between the conventional method (Left) and the proposed method (Right)

Acknowledgement

This paper was supported by Gyeonggi-do and Siheung-city under Industry-University Convergence Campus Outcome Utilization Business, provincial government financial support program, Korea.

References

[1] Zheng, D., Zhao, Y., & Wang, J. (2005). An efficient method of license plate location. *Pattern Recognition Letters*, **26**(15), 2431-2438.

[2] Lee, E. R., Kim, P. K., & Kim, H. J. (1994, November). Automatic recognition of a car license plate using color image processing. *In Image Processing, 1994. Proceedings*. ICIP-94., IEEE International Conference (Vol. **2**, pp. 301-305). IEEE.

[3] Anagnostopoulos, C. N. E., Anagnostopoulos, I. E., Loumos, V., & Kayafas, E. (2006). A license plate-recognition algorithm for intelligent transportation system applications. *IEEE Transactions on Intelligent transportation systems*, **7**(3), 377-392.

[4] Gilly, D., & Raimond, K. (2013). A survey on license plate recognition systems. *International Journal of Computer Applications*, **61**(6).

[5] Ktata, S., & Benzarti, F. (2012, March). License plate detection using mathematical morphology. *In Sciences of Electronics, Technologies of Information and Telecommunications (SETIT)*, 2012 6th International Conference on (pp. 735-739). IEEE.

Self-Adaptive Correction of Heading Direction in Stair Climbing for Tracked Mobile Robots Using Visual Servoing Approach

Peng Ji[1], Aiguo Song[1*], Zimo Song[1], Yuqing Liu[2], Guohua Jiang[2] and Guopu Zhao[3]

[1]School of Instrument Science and Engineering, Southeast University, Nanjing, China
[2]National Key Laboratory of Human Factors Engineering, China Astronaut Research and Training Center, China
[3]Jiangsu Siming Engineering Machinery Co. Ltd., Yangzhou, China

a.g.song@seu.edu.cn

Abstract. In this paper, we describe a heading direction correction algorithm for a tracked mobile robot. To save hardware resources as far as possible, the mobile robot's wrist camera is used as the only sensor, which is rotated to face stairs. An ensemble heading deviation detector is proposed to help the mobile robot correct its heading direction. To improve the generalization ability, a multi-scale Gabor filter is used to process the input image previously. Final deviation result is acquired by applying the majority vote strategy on all the classifiers' results. The experimental results show that our detector is able to enable the mobile robot to correct its heading direction adaptively while it is climbing the stairs.

1. Introduction

In the field of indoor mobile robots, climbing stairs is a basic skill for mobile robots. Climbing stairs commonly include alignment stage, climbing stage and landing stage. In this paper, we focus on the posture correction of mobile robots in the climbing stage. At present, there are not many literatures about the heading correction in climbing stairs for mobile robots. For example, fusing rotational velocity measurements from 3-axial gyroscope and measurements of the stair edges acquired with an onboard camera, authors in reference[1] estimated the quaternion-based attitude by an extended Kalman filter and proposed a centering- and a heading-control control model to guide a tracked mobile robot upstairs by the estimates. The authors in reference [2] proposed a stair climbing algorithm based a Kinect sensor for a tracked mobile robot. This algorithm is able to make the robot have the ability to align to the stairs, keep balance while climbing and land on the stair platform. In the climbing stage, the mobile robot correct its heading direction by the difference of two blocks from the depth image of stairs. A Kinect depth senor is the only equipment needed for all the stages. The authors in reference [3] also used a Kinect senor to correct the attitude of a mobile robot in climbing ramp. The reference [4] describes a tracked mobile robot, which takes a horizontal and a vertical LASER range finder to detect the staircase autonomously, correct the initial posture, climb the stairs and land on the stair platform.

To minimize the number of sensors on the mobile robot, in this paper, we intend to use monocular vision to achieve adaptive correction of the heading direction for the mobile robot in climbing stairs. To this end, we take the heading direction deviation detection problem as a classification problem, and therefore an ensemble heading deviation detector is proposed in this paper to predict whether the mobile robot's heading direction has deviated. According to the predicting results, the mobile robot takes corresponding actions to correct its heading direction.

2. Overview of the System

The images used in this paper are captured by the wrist camera of a self-designed mobile robot. This mobile robot is a kind of tracked mobile robot, which is used for reconnaissance detection [5]. Before self-adaptive correcting the heading direction, the mobile robot rotates the wrist camera to face the stairs, as shown in Figure 1.

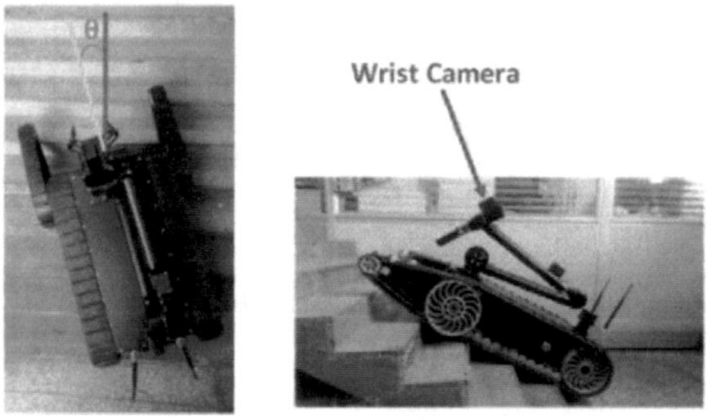

Figure 1. Mobile robot for experiments

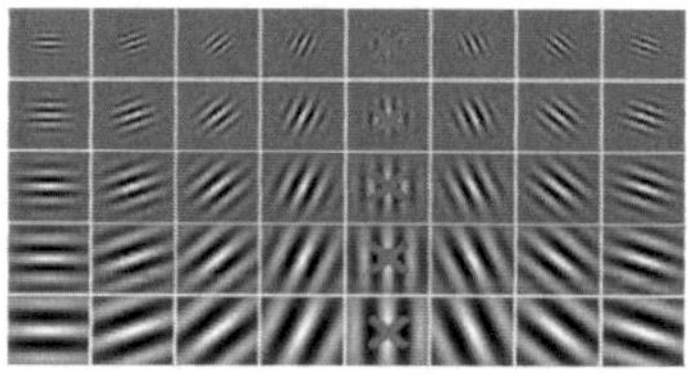

Figure 2. Gabor filter bank

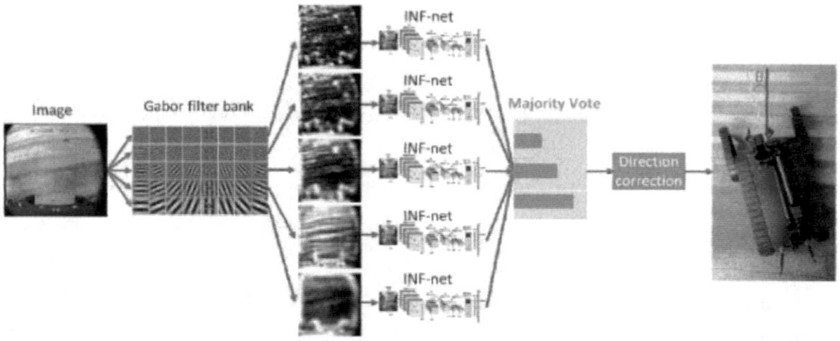

Figure 3. Structure of ensemble heading deviation detector

Our proposed ensemble heading deviation detector consists of three phases, image texture information extraction, training and predicting of ensemble heading deviation detector. Figure 3 shows the overall block diagram of our proposed ensemble heading deviation detector in this paper.

3. Proposed Method

3.1 Multi-scale texture feature extraction

To eliminate the interference of vertical segments in the image, a multi-scale Gabor filter (five scales at seven directions, not including 90 degree) is used to extract the image texture features, as shown in Figure 2. The impulse response of the Gabor filter is defined as a sine wave (2D Gabor filter is a sinusoidal plane wave) multiplied by the Gauss function [6]. The complex form expression of the 2D Gabor filter is shown below:

$$G(x,y) = \frac{1}{2\pi * \sigma^2} exp(-\frac{x^2 + \gamma^2 y^2}{2\sigma^2}) exp(i(2\pi \frac{x}{\lambda} + \psi))$$
$$x = x\cos\theta + y\sin\theta \qquad (1)$$
$$y = -x\sin\theta + y\cos\theta$$

where λ is the wavelength of sine wave, θ represents the direction of parallel wavelet of Gabor wavelet, ψ is phase offset, γ is the aspect ratio of Gabor wavelet, which determines the shape of the Gabor wavelet, σ is standard deviation of the Gaussian function.

3.2 Ensemble heading deviation detector

In this paper, our ensemble heading deviation detector adopts the homogeneous ensemble method, and a variant of Convolutional Neural Network, INF-net, is taken as the basal classifier. The structure of INF-net is shown in Figure 4. The activation function of each layer is ReLU function, and the pooling layer takes max-pooling strategy. The output layer contains three neurons, which represent left deviation, right deviation and no deviating respectively. The INF-net can also be seen as a special case of SPP-net [7] where its pyramid layer number is one.

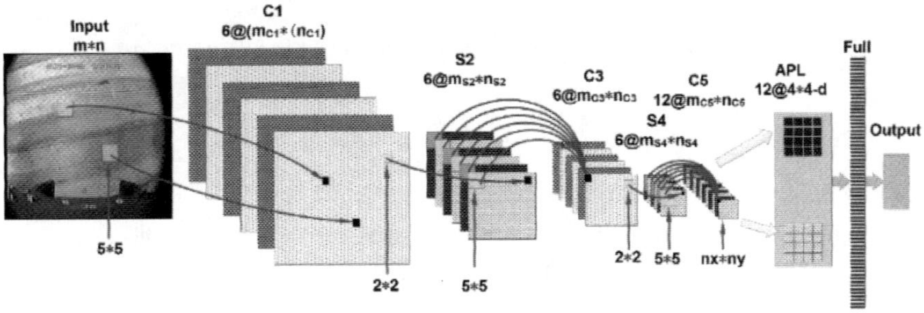

Figure 4. INF-net

In this paper, to get the sample set of our ensemble heading deviation detector, the mobile robot is controlled to climb stairs. After having captured the samples of left deviation, right deviation and no deviating, corresponding labels are given to them. Considering the samples balance, a total of 4850 frames (60×60) is collected. Partial frames are shown in Figure 5. We extract 1/5 frames from those frames randomly as the test sample set, and the remaining frames as the training sample set.

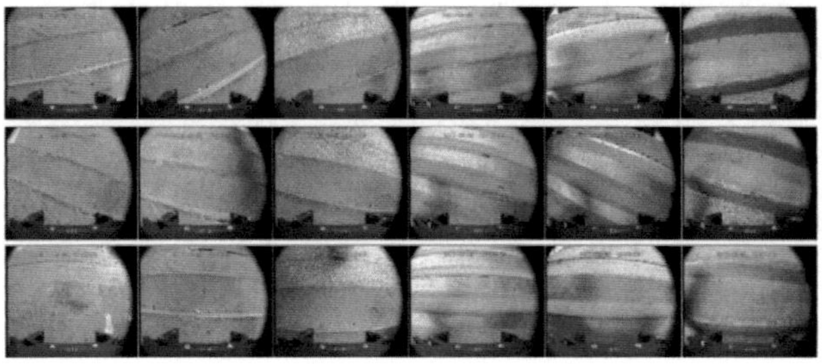

Figure 5. Partial frames

3.3 Predicting and correcting

Heading deviation angle θ is defined as the angle between the robot's heading direction and the centerline of staircase. To improve the generalization ability of the ensemble heading deviation detector, a multi-scale Gabor filter banks is used to process input image previously. Before this preprocessing, the images is converted to grayscale image firstly. At each scale a frame of texture feature image is calculated by summing and normalizing all directions of the texture image. The formula is as follows:

$$F_i = \frac{\sum_{m=1}^{M} f_i^m}{M} \tag{2}$$

where M represents the number of filter directions, f_i^m represents the m-th texture feature image at i-th scale, F_i is the texture feature image at i-th scale. Final deviation result is acquired by applying the majority vote strategy on all the deviation results predicted by each classifier.

$$Final\ Output = majorityvote(Output_1, \dots Output_N) \tag{3}$$

After the ensemble heading deviation detector outputs the final result, the robot correct the heading direction according to this final result. Specifically, the robot correct its heading direction according to the formula (4). Where, $action$ represents the executable action of the mobile robot.

$$action = \begin{cases} right\ forward, & \theta = 1 \\ forward, & \theta = 0 \\ left\ forward, & \theta = -1 \end{cases}, \text{ where } \theta = \begin{cases} -1, & left\ deflection \\ 0, & no\ deflecting \\ 1, & right\ deflection \end{cases} \tag{4}$$

4. Experiments

4.1 k-fold cross validation and comparison experiment

To eliminate the interference of vertical segments in the image, we used a Gabor filter (five scales at seven directions, not including 90 degree) to preprocess the input image. To validate the performance of the detector, we used the 10-fold cross validation method in the testing experiment. The learning rate of each INF-net at five scales was set to 1, the iterations was set to 10, the dropout rate was set to 5% and the kernel size of Gabor filter was all set to 10. The Figure 6(a) shows the training error curve of INF-net at five scales. From the Figure 6(a), we find that the training error decreases rapidly with the increase of the iterations. At the end of each iteration, our proposed algorithm was tested on the test set. The blue line in Figure 6(b) shows the accuracy of our proposed algorithm. In order to verify the performance of our proposed algorithm, we compare our proposed algorithm with the traditional algorithm (single scale gray feature and single classifier). The red line in Figure 6(b) shows the accuracy of the traditional algorithm.

As can be seen from the above figures, the recognition rate of our detector can reach up to 99.67%. From the sixth iteration, the recognition rate has a slight decrease, which may be due to the over-fitting. Overall, our proposed algorithm (multi-scale texture feature extraction + ensemble classifier) has a better performance than the traditional algorithm (single scale gray feature + single classifier). In order to achieve the best results, the five INF-nets was used as the basis classifier of the ensemble heading deviation detector when the sixth iteration was finished.

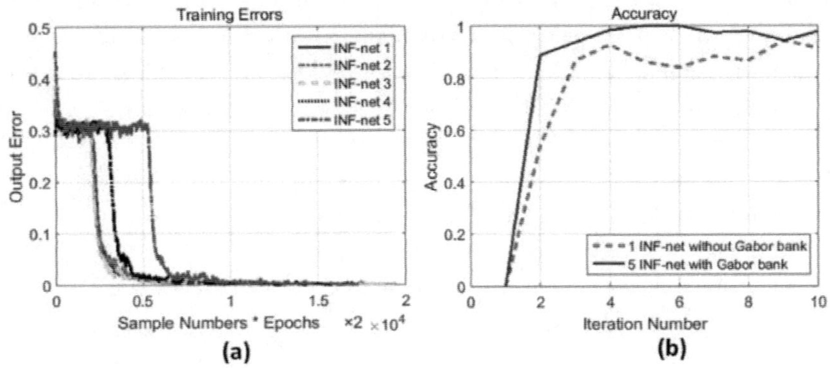

Figure 6. Training errors and accuracy rates

4.2 Artificial disturbance

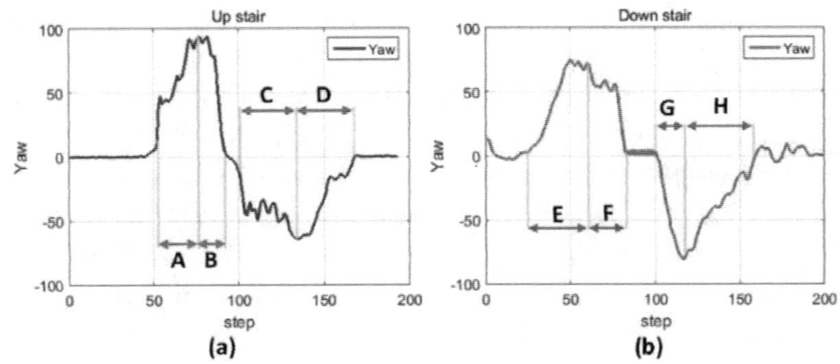

Figure 7. Heading direction correction with artificial disturbance

To verify the correction ability, artificial disturbances were added while the mobile robot was climbing the stairs (ascending-stair and descending-stair). By recording the heading directions, we got the process of our detector correcting the heading direction. Figure 7(a) shows the changes of the heading direction while the mobile robot climbing an ascending-stair, where section A and C are the changes of heading direction while the robot got an artificial disturbance and section B and D are the changes of heading direction while the mobile robot correcting the heading direction. Figure 7(b) shows the case of climbing a descending-stair, section E and G are the artificial disturbance phase and section F and H are the correcting phase. As can be seen from the Figure 7, our proposed algorithm is able to enable the mobile robot correct its heading direction even if it is disturbed by an artificial disturbance, so as to ensure the mobile robot's course has no deviating.

5. Conclusion

In this paper, an ensemble heading deviation detector is proposed, and this detector is applied to adaptively correct the heading direction for the mobile robot in climbing stairs. To improve the generalization ability, we combine the multi-scale Gabor filter with a majority vote strategy. The comparison experimental result shows that our proposed algorithm has a better performance than the traditional algorithm. In addition, an artificial disturbance experimental results show that our algorithm has a good ability to enable the mobile robot to correct its heading direction even if it is disturbed by an artificial disturbance while it is climbing stairs (ascending-stair and descending-stair).

Acknowledgements

This work was supported by National Natural Science Foundation of China (No.61325018, No.61272379), the Defense Industrial Technology Development Program under Grants number (B1720132001), and Natural Science Foundation of Jiangsu Province (BK20141284).

References

[1] Mourikis A.I., Trawny N., Roumeliotis S.I., Helmick D.M., Matthies L. 2007 Autonomous stair climbing for tracked vehicles International Journal of Robotics Research vol 26 pp. 737-758

[2] Li I.H., Wang W.Y., Tseng C.K. 2014 A Kinect-sensor-based Tracked Robot for Exploring and Climbing Stairs International Journal of Advanced Robotic Systems vol 11

[3] Li I.H., Wang W.Y., Chien Y.H., Fang N.H. 2013 Autonomous Ramp Detection and Climbing Systems for Tracked Robot Using Kinect Sensor International Journal of Fuzzy Systems vol 15 pp. 452-459

[4] Mihankhah E., Kalantari A., Aboosaeedan E., Taghirad H.D., Moosavian S. 2008 Autonomous Staircase Detection and Stair Climbing for a Tracked Mobile Robot using Fuzzy Controller Proc. IEEE INTERNATIONAL CONFERENCE ON ROBOTICS AND BIOMIMETICS pp. 1980-1985

[5] Qian K., Song A.G., Bao J.T., Zhang H.T. 2012 Small Teleoperated Robot for Nuclear Radiation and Chem ical Leak Detection International Journal of Advanced Robotic Systems vol 9

[6] Haghighat M., Zonouz S., Abdel-Mottaleb M. 2015 CloudID: Trustworthy cloud-based and cross-enter prise biometric identification Expert Systems with Applications vol 42 pp. 7905-7916

[7] He K.M., Zhang X.Y., Ren S.Q., Sun J. 2015 Spatial Pyramid Pooling in Deep Convolutional Networks for Visual Recognition IEEE Transactions on Pattern Analysis and Machine Intelligence vol 37 pp. 1904-1916

IOP Conf. Series: Journal of Physics: Conf. Series **806** (2017) 012006 doi:10.1088/1742-6596/806/1/012006

Study on Underwater Image Denoising Algorithm Based on Wavelet Transform

Sun Jian[1]**, Wang Wen**[1]

[1]Harbin Engineering University, Heilongjiang, Harbin, 150001, China;

540010224@qq.com

Abstract. This paper analyzes the application of MATLAB in underwater image processing, the transmission characteristics of the underwater laser light signal and the kinds of underwater noise has been described, the common noise suppression algorithm: Wiener filter, median filter, average filter algorithm is brought out. Then the advantages and disadvantages of each algorithm in image sharpness and edge protection areas have been compared. A hybrid filter algorithm based on wavelet transform has been proposed which can be used for Color Image Denoising. At last the PSNR and NMSE of each algorithm has been given out, which compares the ability to de-noising

1. Introduction

Underwater laser imaging technology plays an important part in the exploration of underwater information such as detecting underwater information. However, due to the special water medium, the image will be subject to noise pollution so that the resolution will be decreased. A great deal of manpower and material have been invested in researches of underwater laser ranging, imaging and other fields by many countries such as The United States, the former Soviet Union , and breakthrough has been made in some key directions [1].

There's difference between the characteristics of water and air that laser transports in, which brings great difficulties in underwater target detection using laser. Mainly in the following two points [2]:

1. The effect of laser ranging accuracy and imaging results due to the backscattering of beam by sea water.

2. The limition of the target detection distance due to the strong attenuation of light waves by sea water.

In recent years, a lot of researches on the backscattering and attenuation of laser in seawater have been done at home and abroad, which bring great breakthroughs in underwater laser detection imaging technology, detection distance and image quality[2]. Among them the scattering model of underwater laser has been described in reference [1] which improves the median filter, but the improvement of image processing effect is not obvious. The Wavelet Threshold Denoising has been introduced in reference [2], but the advantages of the algorithm have not been reflected.A variety of commonly used image processing algorithms have been described in reference [3] which simulate them also, but the relevant evaluation parameters of noise canceling have not been given.

In this paper, the transmission characteristics of laser underwater are discussed first, the attenuation of the laser under water is analyzed and the scattering model is calculated. Furthermore, the noise of underwater laser imaging is analyzed, the common noise suppression algorithm and their advantages and disadvantages are given. At last, a filtering algorithm based on wavelet transform is proposed and simulated, and the Peak Signal to Noise Ratio, Normalized Mean Square Error and operation hours of each algorithm are calculated, which show the advantage of wavelet threshold denoising.

2. Analysis of Underwater Laser Propagation

The attenuation and scattering characteristics of laser signal in water have been analyzed in his chapter.

2.1 The attenuation of laser signal in water

The attenuation coefficient of laser is relatively large in the short wavelength and long wavelength region, but small in the blue-green region. That's what we called Seawater Optical Window as usual, which played an important role in the development of ocean optics. So far, most of the lasers used in ocean lasers are blue-green lasers, so that the laser attenuation in the water is the smallest, which can reach the deeper Detection distance [3].

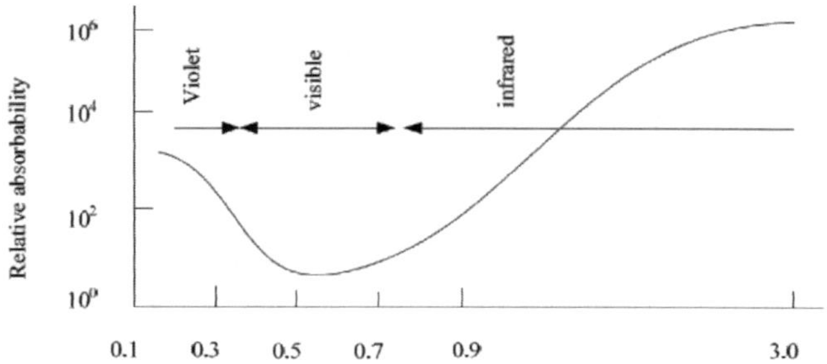

Figure 1 relative absorption coefficients of light waves at different wavelength

It can be seen from the trend diagram of the relative absorption coefficient of wavelength in Figure1 that the attenuation of ultraviolet and infrared light band in water is great large so that they can't be used in water. However, the attenuation of blue-green light in the visible light band is minim, for example, the attenuation lengths of the light waves in 0.4900μm and 0.6943μm are 11m and 2m. It shows that the transmission performance of blue light in water is better than red light.

2.2 The scattering of aqueous media

The reduction of signal to noise ratio of underwater laser imaging receiving system is the major effect of the scattered light which will bring a strong background noise. As the imaging distance increases, the screen contrast reduced and the image contrast decreases so that underwater optical imaging becomes more difficult[4]. What's more, because of the particles, plankton in seawater and the flow of water, the noise will be generated inevitably for the underwater images which causes the decrease of imaging quality further[5].

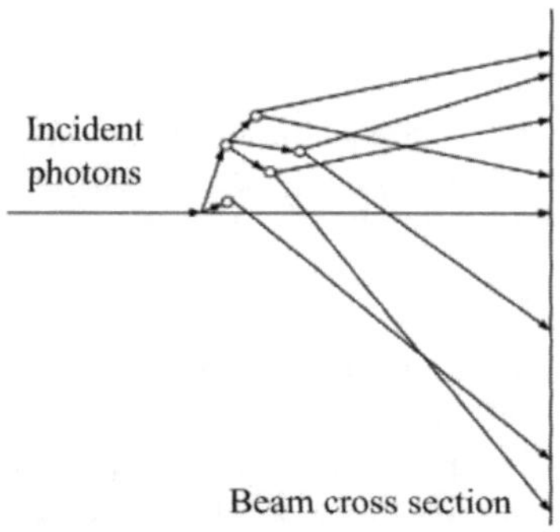

Figure 2 forward scattergram of water molecules to light

2.2.1 The forward scattering. As we can be seen from Figure 2, the forward scattering increases the laser transmission distance significantly, and the farther the transmission distance, the greater the role of forward scattered light, which is advantageous to underwater illumination, but disadvantageous to underwater beam scanning and imaging because it will degrades resolution and target background contrast.

2.2.2 Backscattering. The scattering of light in the direction opposite to the propagation direction is called backscattering. Strong backscattered light will saturate the receiver without receiving any useful information, that's why the main research tries to overcome the impact of backscatter in underwater ranging, imaging and other applications. There are the following methods of overcoming the effects of backscattering: Using a suitable filter and analyzer; Separating emission sources and receivers as far as possible. Using the distance gating technology is the most effective method still now [6].

3. Underwater Image Denoising Algorithm

Digital image processing is generally from the beginning of image filtering so it's necessary to understand the type and nature of the noise.

3.1 Classification of image noise

One of the important factors that cause the image resolution to decrease is the noise which from the image acquisition and transmission process. It is necessary to know the type and nature of noise because the processing of the image is generally started with image filtering. From the category point of view, the main noise has the following:

(1) Gaussian noise: It is the most common kind of noise which the various unstable factors of imaging system are often expressed through.

(2) Salt and pepper noise: The noise that shows larger and smaller gray scale values in a certain pixel or a region is what we called salt and pepper noise.

(3) Fixed mode noise: It is due to imperfections in the system devices. Because it does not have randomness, we also call it "The clutter".

The actual image signal is often affected by correlated noise, signal-related noise, and various other distributed noises. Therefore, it is impossible to strictly determine and calculate noise.

3.2 Common denoising methods

Here are some common denoising algorithms: Wiener filter, median filter, average filter algorithm

3.2.1 The Adaptive Wiener Filter. The Adaptive Filter can adjust the current filter parameters by the filter parameters that can be obtained empirically which make it adapt to the statistical properties of unknown or time-varying of signals and noise [7]. The smoothing effect of the filter is small when the local variance is large and stronger when the local variance is small. The ultimate goal of the Adaptive Wiener Filter is to minimize Mean Square Error $\sigma^2=E[(f(x,y)-F(x,y))^2]$ of the restoring image $f(x,y)$ and the original image $F(x,y)$. This method is useful for preserving edges and other high-frequency parts of an image but its calculation is large.

3.2.2 The Median Filtering. The Median Filtering is a nonlinear signal processing method which can overcome the image detail blur caused by linear filtering under certain conditions and effective for filtering pulse noise and image scanning noise. The median filter not only removes noise but also protects the edges of the image which can obtain a more satisfactory recovery effect. It's convenient that the Statistical Properties of Images is not necessary. However, the Median Filtering is not suitable for processing the image with a lot of points, lines, spire details. The Median Filtering can be represented by the following formula:

$$\hat{f}(x,y) = \underset{(s,t)\in S_{xy}}{Med}\left\{g(s,t)\right\} \tag{1}$$

3.2.3 The Average Filter. Here are two average denoising filtering algorithms: The Arithmetic Average Filter; The Geometric Average Filter.

(1) The Arithmetic Average Filter

Arithmetic Average Filter is a simple average filter. Sxy represents a coordinate group of the rectangular image window that the center is at (*x*,*y*) and the dimension is m×n. The arithmetic mean process is to calculate the average of the interfered images *g*(*x*,*y*) in the defined area Sxy. The value of the restored image $\hat{f}(x, y)$ at any point (*x*,*y*) is the arithmetic average calculated by the pixels of the area defined by S_{xy}:

$$\hat{f}(x,y)=\frac{1}{mn}\sum_{(s,t)\in S_{xy}} g(s,t) \tag{2}$$

This can be achieved by a convolutional template whose coefficient is 1/mn.

(2) The Geometric Average Filter

The image restored with the geometric average filter is given by the following expression:

$$\hat{f}(x,y)=\frac{mn}{\sum_{(s,t)\in S_{xy}} \frac{1}{g(s,t)}} \tag{3}$$

The harmonic averaging filter is more effective for salt noise and gaussian noise, but not suitable for pepper noise.

3.2.4 The wavelet denoising. The basic methods of wavelet denoising are: The wavelet transform modulus maxima denoising; The wavelet transform inter-scale correlation denoising; The nonlinear wavelet threshold denoising.

(1) The wavelet transform modulus maxima denoising

The wavelet transform modulus maxima denoising method is based on the different propagation characteristics of signal and noise at each scale of wavelet transform, removing the modulus maxima point generated by noise, preserving the modulus of the signal corresponding to the maximum point, and then reconstructing the wavelet coefficients by the residual modulus maxima points, restoring the signal at last.

(2) The wavelet transform inter-scale correlation denoising

There is a correlation between the wavelet coefficients corresponding to the image features in different scale space (resolution), which we called the correlation between the scales. This correlation is fixed in the wavelet transform decomposition process which reflects the multi-scale nature of space. In the same scale space, the important wavelet coefficients are concentrated in some areas, and this correlation is called intra-scale correlation. According to the above characteristics of wavelet transform between signal and noise at different scales, it is possible to enhance the signal and suppress the noise by multiplying the wavelet coefficients of adjacent scales directly.

(3) The nonlinear wavelet threshold denoising

The wavelet transform has strong data decorrelation, which can make the energy of the signal concentrate in a small number of wavelet coefficients in the wavelet domain, while the noise is distributed in the entire wavelet domain, corresponding to a large number of small wavelet coefficients. The amplitude of the wavelet coefficient of the signal is larger than the amplitude of the wavelet coefficients of the noise after wavelet decomposition so that we can keep the wavelet coefficients of the signal, leaving most of the noise wavelet coefficients reduced to zero by the threshold method.

The qualitative comparison of the three denoising methods is shown in Table 1.

Table 1 The qualitative comparison of the three denoising methods

denoising methods	modulus maxima	inter-scale correlation	threshold denoising
Calculation	Large	Relatively large	Small
Stability	Stable	Relatively stable	Depend on SNR
Effection	Relativel Good	Good	Good
Range	Low SNR	High SNR	High SNR

3.3 Color Image Denoising Algorithm Based on Wavelet Threshold

There is different denoising ability for different noise, and their respective advantages and disadvantages for the common denoising methods. For example, The Median Filter has better denoising ability to impulse noise, but less effect on Gaussian noise. The Average Filter has better denoising ability to Gaussian noise, but not conducive to protect the edge of image so that the image becomes indistinct. The Wiener filter is useful for preserving edges and other high-frequency parts of the image, but computationally expensive

The following proposes a filtering algorithm based on the wavelet threshold filter, which can not only remove the image of Gaussian noise and salt and pepper noise better, but also preserve the image clarity better. The algorithm is based on the wavelet threshold filtering algorithm and use the median filter combined with mean filter combination, which makes the threshold selection of the algorithm easy. After wavelet threshold denoising, the Fourier transform and Butterworth high-pass filter are used to preserve the edges of the image.

The threshold functions which is divided into hard threshold function and soft threshold function in the threshold denoising algorithm, represent different processing strategies and estimation methods for the wavelet coefficients that exceed and fall below the threshold.

The most important thing in the threshold denoising algorithm is how to choose threshold and threshold function because the noise still exists in the denoised image if the threshold is large, and important image features will be filtered out if the threshold is small. Intuitively speaking, for a given wavelet coefficient, the greater the noise, the large the threshold. So most of the threshold selection process is calculating a threshold according to the statistical characteristics of the wavelet coefficients. Therefore most of the threshold selection process is for a set of wavelet coefficients, which means a threshold is calculated according to the statistical characteristics of the group of wavelet coefficients.

The flow chart of the hybrid noise filtering algorithm is shown in Figure 3.

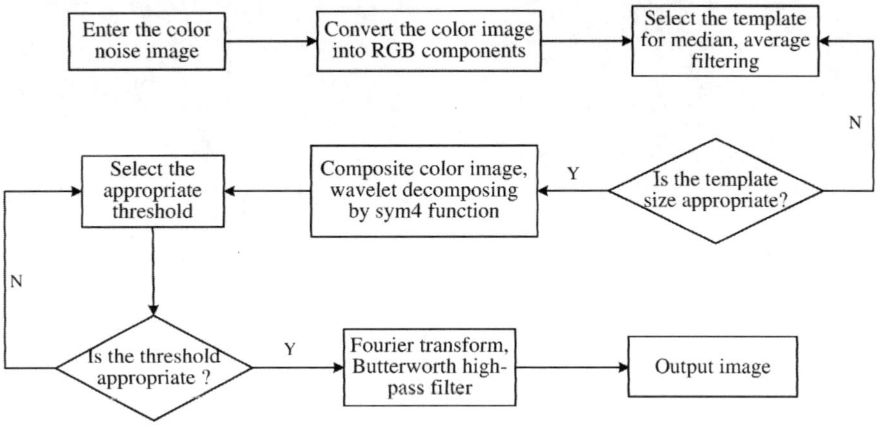

Figure 3 The flow chart of the hybrid noise filtering algorithm

4. Simulation Experiment and Result Analysis

In this chapter, the related algorithms introduced in the previous chapter are simulated and the PSNR and NMSE of each algorithm has been given out, which compares the ability to de-noising.

4.1 Quality Evaluation of Noise Canceling

Some common image fidelity measurement parameters are given: Mean Absolute Error (MAE), Mean Square Error(MSE), Normalized Mean Square Error(NMSE), Signal to Noise Ratio(SNR), Peak Signal to Noise Ratio(PSNR) [8].

The SNR above 40dB shows excellent image quality close to the original image, 30~40dB usually indicates that the image quality is good that distortion can be perceived but acceptable, 20~30dB shows poor image quality; the SNR below 20dB means the image is unacceptable [9].

The PSNR and NMSE are used as the objective evaluation index in this paper. The following describes its calculation method.

$$PSNR=10\cdot\log\frac{255^2}{\frac{1}{MN}\sum_{i=0}^{M-1}\sum_{j=0}^{N-1}\left(f(i,j)-p(i,j)\right)^2} \tag{4}$$

$$NMSE=\frac{\sum_{i=0}^{M-1}\sum_{j=0}^{N-1}\left(f(i,j)-p(i,j)\right)^2}{\sum_{i=0}^{M-1}\sum_{j=0}^{N-1}f(i,j)^2} \tag{5}$$

In the formula, M and N mean the height and width of the image respectively; $f(i,j)$ mean the pixel value of the original image; $p(i,j)$ means the pixel value of the filtered image. The larger the SNR, the better the noise smoothing effect, while the smaller the MSE, the less the edge of image loss [10].

4.2 Experimental simulation and result analysis

In this paper, the MATLAB is used to analyze the image processing. The filtering effects of some commonly used filtering functions are compared, Figure 4 shows the original noise image, Figure 5 to Figure 9 show the images processed by the usual filtering algorithm, and Figure 10 to Figure 12 show the images processed by the three wavelet algorithms.

Figure 4 original noise image

Figure 5 3×3 median filtering

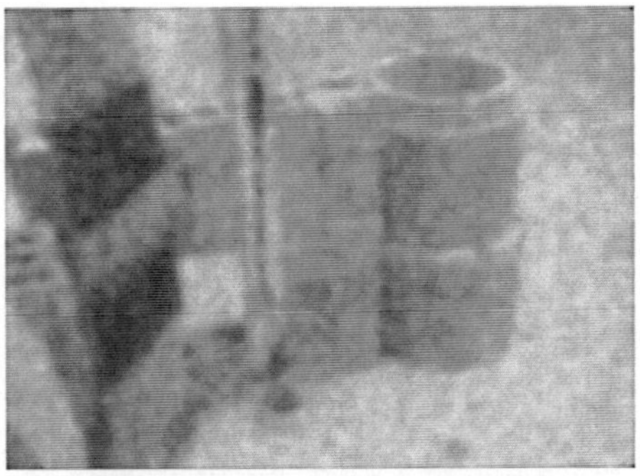

Figure 6 5×5 median filtering

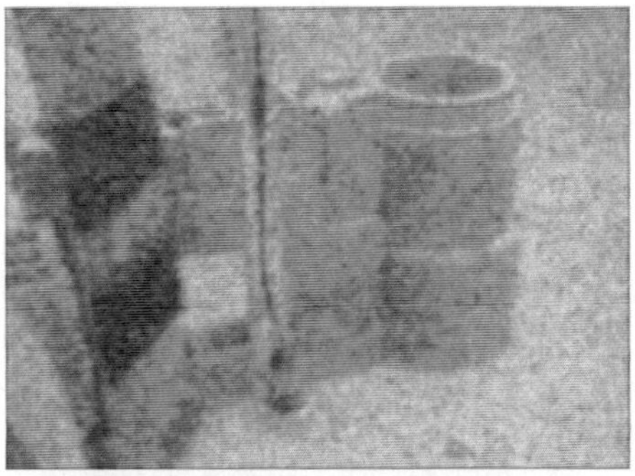

Figure 7 3×3 average filtering

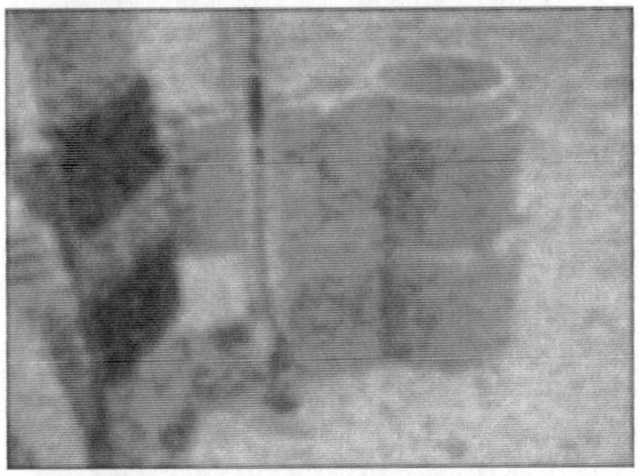

Figure 8 5×5 average filtering

Figure 9 Wiener filtering

Figure 10 wavelet threshold filtering

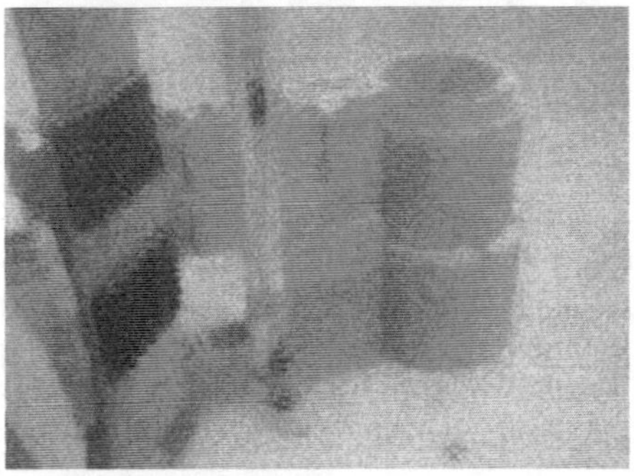

Figure 11 modulus maxima filtering

Figure 12 scale correlation filtering

By contrast, the image can reflect the smaller median filter and the mean filtered image, but the noise removal is less, and the larger template can remove the noise better, but it will make the image edge blurred. In addition, compared to other commonly used algorithm, the Wiener filter's image clarity is better, but the pepper noise removal capacity is weak. Compared with the three wavelet denoising algorithms, the image filtered by the modulus maxima is blurred and the image information is distorted. However, the image processed by wavelet threshold filtering and scale correlation filtering is clearer, and the edge of image can be well protected.

Table 2 Comparison of PSNR and NMSE in Common Algorithm

Evaluation	PSNR	NMSE
3×3 median	33.0417	0.0308
5×5 median	33.8768	0.0333
3×3 average	32.0843	0.0302
5×5 average	31.7273	0.0356
Wiener	34.8971	0.0209

Table 3 Comparison of PSNR and NMSE in Wavelet filtering

Evaluation	PSNR	NMSE
wavelet threshold	37.7064	0.0160
modulus maxima	36.0106	0.0199
scale correlation	37.3846	0.0152

Table 2 shows that the Wiener filter has the highest PSNR value, the best noise removal effect, and the lowest value of NMSE, and the better protection of the image edge in the commonly used image processing algorithms. Table 3 shows that PSNR is slightly lower than the other two algorithms and NMSE is slightly higher for the three wavelet denoising algorithms; the PSNR and NMSE of the wavelet threshold filter and the scale correlation filter are close.

Table 4 algorithm program running time

Algorithm	wavelet threshold	modulus maxima	scale correlation
Running time	2.15s	4.82s	2.53s

Table 4 shows that the time spent of filtering the modulus maxima is close to that of threshold filtering, but the scale-dependent filtering is computationally intensive and takes a long time to run. Comparing the PSNR value, RMS value and running time of the three algorithms, when the threshold is more appropriate, the wavelet threshold filter algorithm has a higher SNR and a shorter running time, which is superior to the other two algorithms.

5. Conclusions

The paper introduces the transmission characteristics of underwater laser signals and summarized the types of underwater imaging noise. In this paper, commonly used noise suppression algorithms were elaborated, including wiener filtering, median filtering, mean filtering. Furthermore the advantages and disadvantages of each algorithm are analyzed. Then based on the proposed three wavelet denoising algorithms, the peak signal SNR and normalized mean square error of each algorithm are calculated, and the running time of each algorithm are also compared. Finally, it is found that the wavelet threshold filtering algorithm not only removes noise effectively, increases the SNR of the image and protects the edge of the image better, but also has shorter running time than the other two algorithms.

IOP Conf. Series: Journal of Physics: Conf. Series **806** (2017) 012006 doi:10.1088/1742-6596/806/1/012006

References

[1] Jaffe J S, Moore K D, Mclean J, et al. Underwater Optical Imaging: Status and Prospects. Oceanography, 2001, 14(3):1-3.
[2] R. Garcia, T. Nicosevici and X. Cufi, "On the Way to Solve Lighting Problems in Underwater Imag IEEE OCEANS Conference (OCEANS), Biloxi, Mississipi pp. 1018-1024, 2002.
[3] J. Portilla, M.W. V. Strela, and E. P. Simoncelli, "Image denoising using scale mixtures of Gaussians in the wavelet domain," IEEE Trans. Image Process., vol. 12, no. 11, pp. 1338–1351, Nov. 2003.
[4] H. R. Sheikh and A. C. Bovik, "Image information and visual quality," presented at the IEEE Int. Conf. Acoust., Speech, Signal Processing, May 2004.
[5] He Duo-min, Seet Gerald G.L. Laser gated-ranging for underwater robot vision in turbid waters. Proceedings of SPIE - The International Society for Optical Engineering. 2002, 4546: 11-22
[6] Berechet I, Berginc G. 3D laser imaging for concealed object identification. Proceedings of SPIE - The International Society for Optical Engineering, 2014, 9205: 2050L-15.
[7] Hou W,Gray D J,Weidemann A D, et al. Automated underwater image restoration and retrieval of related optical properties. Geoscience and Remote Sensing Symposium, 2007. IGARSS 2007. IEEE International. 2007:1889-1892.
[8] Tan C S, Sluzek A, Seet G L G, et al. Range Gated Imaging System for Underwater Robotic Vehicle. Oceans. 2007:1-6.
[9] C. S. Tan, G. Seet, A. Sluzek, D. M. He, "A Novel Application of Range Gated ULIS in Near Target Turbid Medium", Journal of Optics and Laser in Engineering, Vol. 43,. 995-1009, 2005.
[10] Sheikh H R, Bovik A C. Image information and visual quality. IEEE Transactions on Image Processing A Publication of the IEEE Signal Processing Society, 2006, 15(2):430-44.

Temporal Classification Error Compensation of Convolutional Neural Network for Traffic Sign Recognition

Seungjong Yoon, Eungtae Kim

Dept. Electronics Engineering, Korea Polytechnic University, 237, Sangidaehak-ro, Siheung-si, Gyeonggi-do, Korea

Abstract. In this paper, we propose the method that classifies the traffic signs by using Convolutional Neural Network(CNN) and compensates the error rate of CNN using the temporal correlation between adjacent successive frames. Instead of applying a conventional CNN architecture with more layers, Temporal Classification Error Compensation(TCEC) is proposed to improve the error rate in the architecture which has less nodes and layers than a conventional CNN. Experimental results show that the complexity of the proposed method could be reduced by 50% compared with that of the conventional CNN with same layers, and the error rate could be improved by about 3%.

1. Introduction

Recognizing traffic signs is a crucial technology for either drivers or automotive vehicles because it can provide lots of necessary information for drivers. Compared to the conventional location-based traffic sign information, recognition through video frames enables drivers to respond more flexibly by collecting real-time information in fast-changing environments. Recognizing traffic signs can be classified using the Convolutional Neural Network (CNN) [1, 2, 3], which is known to perform well in image data among the Artificial Neural Networks (ANN). Considering the characteristics of ANN, the higher number of nodes and layers can help enhance performance.

Beyond certain point, however, more layers and nodes are required in order to enhance the accuracy rate. As a result, as the number of nodes and layers increases, the computational complexity of CNN methods increases exponentially and the real-time processing becomes difficult. In this paper, we propose the Temporal Classification Error Compensation (TCEC), which can enhance the error rate while restricting the number of nodes and layers. This paper proposes two methods: a) classifying traffic signs using CNN b) compensating the error rate by referencing the adjacent successive frames.

First, in Chapter 2 of this paper, we present the related work about conventional method. In Chapter 3, we look into the CNN architecture with the proposed TCEC, and Chapter 4 will introduce the result of a comparison experiment between the proposed method and the conventional CNN methods. Chapter 5 will present the final conclusion.

2. Related Works

Traffic Sign Classification aims to classify the detected traffic signs into specific classes. Traffic signs are designed to be easily recognizable. For example, colors, shapes, and icons are obvious. These features are also used for learning the networks and classification. Feature extraction algorithms such as Histogram of Oriented Gradients(HOG) and Speed-Up Robust features(SURF) are used to extract the features from the images. For classification, Support Vector Machine (SVM), Random Forest, and

Deep Neural Network (DNN) could be used as the classifier. Particularly, CNN, which is one kind of DNN, is attracting attention because it has good performance in object recognition recently.

The Committee of CNNs [4], which performed best in the IJCNN2011 competition [5], learned HOG and HAAR features through CNN. Among the methods other than CNN, there were Random Forest method [6] and hierarchical SVM method [7]. The CNN method has performed better than the human [8]. However, in the architecture, the number of filters used in the three convolutional layers was 100, 150, and 250, respectively, and 200 neurons were used in the fully connected layer. Because of the large number of the nodes and layer, it had large time complexity. For real-time implementation, more researches are needed to maintain good detection performance while adjusting to have appropriate nodes and layers.

3. Proposed Method

3.1. Classification

In the classification stage, the traffic signs extracted from the frames are classified into 43 classes. Figure 1. shows the CNN model that has two convolution layers. The extracted traffic signs are converted to 28 X 28 size. They are then sent to the C_1 Convolution Layer of Figure 1. C_1 consists of 16 of different 5 X 5 size filters. Each filter is convoluted to the input image, and creates the feature maps same as the number of filters. Then, they go through one sub-sampling and ReLu[9]. ReLu Layer helps to solve several issues that can occur in network learning, and can enhance the learning speed by saving the conventional Sigmoid calculation time.

In order to prevent overfitting problems in the learned network, Dropout Layer [10] is added. Dropout is a method of deleting arbitrary nodes that were used in learning in order to prevent overfitting when the data are propagated while they are learned. When a network is overfitted, it can have good performance for the data set that have been used for training, but it shows very poor performance for the test real data set. Dropout can be carried out in a relatively simple manner, and can reduce the network complexity because the Dropout Layer reduces the number of nodes. The Dropout Layer deletes same number of arbitrarily nodes as the dropout rate. The speed of learning is delayed a little depending on deleting of nodes.

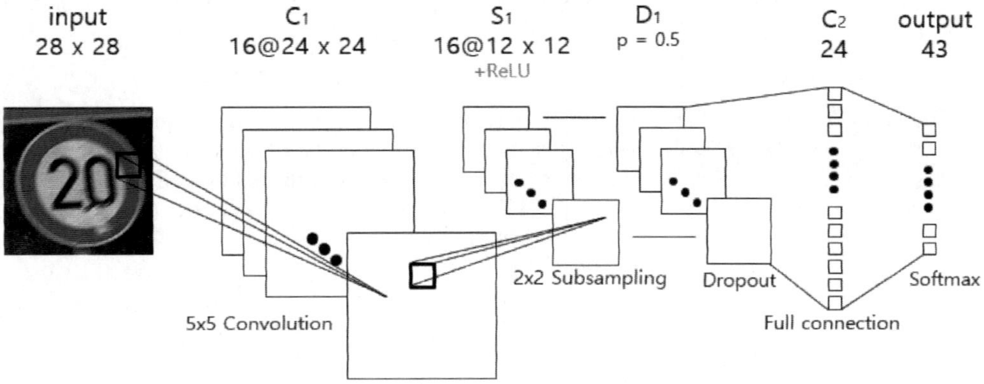

Figure 1. Architecture of the proposed network.

After going through the Dropout Layer, the classification is completed through the Full connection and Softmax stages. As shown in Figure 1, the final output value of CNN equals the number of classes, which is 43. The predicted image class, which has been the input source, has the largest class value.

3.2. Temporal Classification Error Compensation (TCEC)

The error rate could be compensated by making adjustment to the results of current frame, referring to the expected results of the N number of adjacent frames that has processed the classification stage.

Making reference and holding the results of N can be done in two ways: a) referring to the past or future frames only and b) referring to the past and future frames simultaneously.

When a burst error occurs in method a), the error rate rather often increases because the adjustment can be made to the results of current frame with the error rate (that has been failed to be predicted) in the boundary surface of the cluster error. On the other hand, when using method b), the error rate is not high while adjustment can be made to other random errors because of the non-dominant errors in the burst error boundary surface. In other words, TCEC is more effective for local random errors than block errors.

When N number of frames are referred in order to compensate the error rate, the predicted results of CNN class are sequentially saved in the buffer with the size of N. After the buffer is filled with N number of results, the predicted result, placed in the center of buffer as the dominant predicted cost, is sent as feedback of the CNN result in order to adjust to the predicted results. As shown in **Figure 2**, expected results of the five classes are placed sequentially, and after all of the buffers are filled, the most dominant predicted results are used for the adjustment. There are total of two results predicted as Class 11, and total of three results predicted as Class 13. Therefore, the class result is compensated using 13, which is the most dominant result among the five different frames. In case there is no dominant result or it overlaps, the current predicted results are maintained. When applying TCEC, it should be noted that the object within the referred frame needs to be the exact same object. This is because referring to different object can lead to an error propagation.

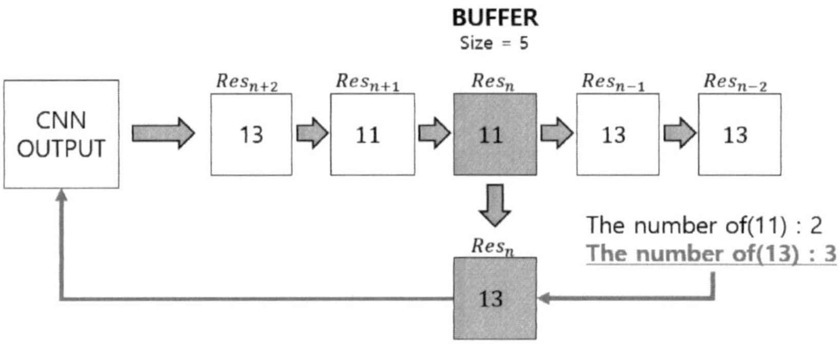

Figure 2. Example of Temporal Classification Error Compensation.

When the (N=5) of buffer of size is used, if there are too many referencing frames, the possibilities of error increases and the temporal correlation with the frame decreases. On the other hand, if very few frames are used as a reference, it can be said that the predicted results of CNN are not dominant enough. Therefore, the number of referred frames must be set considering the error rate and environment where errors occur.

4. Experiment

For the learning of CNN in this experiment, the German Traffic Sign Recognition Benchmark(GTSRB) dataset provided by the Institute for Neural Computation (Institut für Neuroinformatik) in Germany [5]. This includes approximately 40,000 traffic sign learning data and the dataset to test the performance. The dataset includes the images of various sizes and environments.

The 40,000 GTSRB training data were learned after being converted into 28 X 28 grey-scaled image. The number of layers and filters (f_n) were altered for two models to conduct a comparison experiment, and the data were learned for 30 epochs, with an added dropout layer that will prevent overfitting. In this paper, the dropout rate has been set as 0.5.

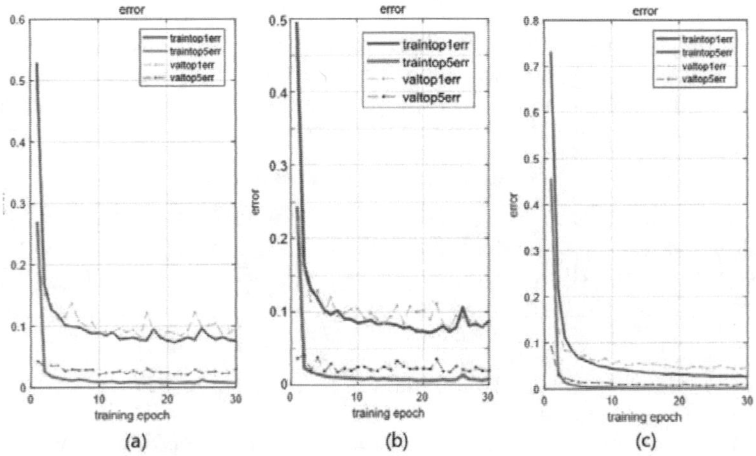

Figure 3. Result of the Learning (a) proposed method without TCEC, (b) 2 layer Conventional CNN Model 1, (c) 3 layer Conventional CNN Model 2

In the experiment, the conventional two models with different numbers of nodes and layers were compared with the proposed method. In Figure 3, (a) shows the learning result of proposed method with 2 layers and no TCEC, (b) is the result of Conventional CNN Model 1 with two layers, and (c) shows the learning result of the Conventional CNN Model 2 with three layers. The solid lines are the error rate for the training data, and the dotted lines are the error rate for the validation data. As it is shown in the graph, the error rate of (c) is almost half of (a). While performance enhances when there are three layers, the complexity almost triples compared to the proposed method. In this experiment, A PC with 3.4 GHz Intel Core i7 CPU, 16 GB memory was used, and the experiment was conducted on MATLAB using matConvnet library.

The video images that were used for the actual test are driving car clip images in Germany and edited for 1~4 seconds. A total of 42 clip sequences were used for the experiment. Figure 4 shows some examples of the footages that were used. They were 1280 × 720 HD resolution videos. While they were 25fps videos, the experiment was done after sampling them into 5fps videos in order to prevent the data from becoming massive. The total number of frames of test video were 749, and there were also the frames including partially captured traffic signs as shown in Figure 5. Among the videos, 296 images were captured under hard condition such as rainy.

Figure 4. Examples of Test data samples

Figure 5. Examples of incomplete signs

Each frame has information on the location of traffic signs and classes, RoI areas are extracted within the frames. The images were input after being converted into grey-scaled images of 28 X 28 size, which is the same as the learned images. The information on classes is used to calculate the error rate. In TCEC, compensation was done after referring two past and future frames, respectively.

Table 1. Results of the experiment.

	Conventional CNN Model 1 with 2 layer $(\mathbf{f_1 = 32, f_2 = 64})$	Conventional CNN Model 2 with 3 layer $(\mathbf{f_1 = 16, f_2 = 32, f_3 = 64})$	Proposed Method without TCEC $(\mathbf{f_1 = 16, f_2 = 24})$	**Proposed Method with TCEC** $(\mathbf{f_1 = 16, f_2 = 24})$
Number of Errors	119	81	122	**89**
Error Rate	0.1589	0.1081	0.1629	**0.1188**

As the results of an experiment are shown in Table 1, while the proposed method without TCEC had higher error rate than the conventional CNN model 1 with two layers, the proposed method with TCEC had lower error rate than it despite the fact that it had more number of nodes. In addition, the error rate of the proposed method with TCEC was similar with that of the conventional CNN model 2 with 3 layer, which has three times more complexity than the proposed method.

$$O\left(\sum_{l=1}^{d} n_{l-1} \cdot s_l^2 \cdot n_l \cdot m_l^2\right) \quad (1)$$

Equation (1) can be used to calculate the time complexity of all layers of CNN. In the equation, l refers to an index number of convolution layer, d is a depth of the convolution layer. n_l is a width of filter, and s_l refers to the filter size. m_l is a size of feature map [11]. The higher the time complexity, the slower the learning speed of the network.

The time complexity of conventional CNN model 1 is 755,712, and that of conventional CNN model 2 is 1,082,368. The time complexity of proposed method is 285,696. As the proposed method uses classified results of each frame in the conventional method, the complexity does not increase. While the results are output after frame 2, the proposed method improved the error rate even with using less number of nodes and layers compared to models 1 and 2, with more than half their complexities.

5. Conclusion

In this paper, a method of Convolutional Neural Network(CNN) with Temporal Classification Error Compensation(TCEC) for recognizing traffic signs in video sequences has been proposed. It was able to enhance the performance of CNN while not even requiring additional system resources. The proposed method is suitable to be used at sections where traffic signs sequentially appear within a frame. Using the proposed TCEC, we were able to decrease the complexity more than halve compared to conventional CNN models while enhancing the error rate by an additional 3%. It is expected that the proposed method can be effectively applied to autonomous vehicle system or advanced driver assistance systems (ADAS).

Acknowledgement

This paper was supported by Gyeonggi-do and Siheung-city under Industry-University Convergence Campus Outcome Utilization Business, provincial government financial support program, Korea.

IOP Conf. Series: Journal of Physics: Conf. Series **806** (2017) 012007 doi:10.1088/1742-6596/806/1/012007

References

[1] LeCun, Y., Bottou, L., Bengio, Y., & Haffner, P. (1998). Gradient-based learning applied to document recognition. *Proceedings of the IEEE*, **86**(11), 2278-2324.

[2] LeCun, Y., Kavukcuoglu, K., & Farabet, C. (2010, May). Convolutional networks and applications in vision. In *ISCAS* (pp. 253-256).

[3] Sermanet, P., Chintala, S., & LeCun, Y. (2012, November). Convolutional neural networks applied to house numbers digit classification. In *Pattern Recognition (ICPR), 2012 21st International Conference* on (pp. 3288-3291). IEEE.

[4] Cireşan, D., Meier, U., Masci, J., & Schmidhuber, J. (2011, July). A committee of neural networks for traffic sign classification. *In Neural Networks (IJCNN), The 2011 International Joint Conference* on (pp. 1918-1921). IEEE.

[5] Stallkamp, J., Schlipsing, M., Salmen, J., & Igel, C. (2011, July). The German traffic sign recognition benchmark: a multi-class classification competition. In *Neural Networks (IJCNN), The 2011 International Joint Conference* on (pp. 1453-1460). IEEE.

[6] Zaklouta, F., Stanciulescu, B., & Hamdoun, O. (2011, July). Traffic sign classification using kd trees and random forests. *In Neural Networks (IJCNN), The 2011 International Joint Conference* on (pp. 2151-2155). IEEE.

[7] Wang, G., Ren, G., Wu, Z., Zhao, Y., & Jiang, L. (2013, August). A hierarchical method for traffic sign classification with support vector machines. *In Neural Networks (IJCNN), The 2013 International Joint Conference* on (pp. 1-6). IEEE.

[8] Stallkamp, J., Schlipsing, M., Salmen, J., & Igel, C. (2012). Man vs. computer: Benchmarking machine learning algorithms for traffic sign recognition. *Neural networks*, **32**, 323-332.

[9] Krizhevsky, A., Sutskever, I., & Hinton, G. E. (2012). Imagenet classification with deep convolutional neural networks. In *Advances in neural information processing systems* (pp. 1097-1105).

[10] Srivastava, N., Hinton, G. E., Krizhevsky, A., Sutskever, I., & Salakhutdinov, R. (2014). Dropout: a simple way to prevent neural networks from overfitting.*Journal of Machine Learning Research*, **15**(1), 1929-1958.Another reference

[11] He, K., & Sun, J. (2015). Convolutional neural networks at constrained time cost. In *Proceedings of the IEEE Conference on Computer Vision and Pattern Recognition* (pp. 5353-5360).

Proposal: A Hybrid Dictionary Modelling Approach for Malay Tweet Normalization

Nor Azlizawati Binti Muhamad, Norisma Idris and Mohammad Arshi Saloot

University of Malaya, 50603, Kuala Lumpur, Malaysia.

azlizamuhamad@gmail.com

Abstract. Malay Twitter message presents a special deviation from the original language. Malay Tweet widely used currently by Twitter users, especially at Malaya archipelago. Thus, it is important to make a normalization system which can translated Malay Tweet language into the standard Malay language. Some researchers have conducted in natural language processing which mainly focuses on normalizing English Twitter messages, while few studies have been done for normalize Malay Tweets. This paper proposes an approach to normalize Malay Twitter messages based on hybrid dictionary modelling methods. This approach normalizes noisy Malay twitter messages such as colloquially language, novel words, and interjections into standard Malay language. This research will be used Language Model and N-grams model.

1. Introduction

Nowadays, there are so many social media or social networking for connecting people such as Facebook, Istagram, Whatsapp, WeChat, Twitter and so on. Dasselaar, (2016) mentioned that social media as an excellent mechanism for two-way communication[1]. Twitter is one of a free online social networking has above 500 million subscribers[2-3], it is also known as microblogging service that enables registered members to post a short note in 140 characters, beside that it also may contain photos, videos, links called as tweets, but to those unregistered Twitter only can read a tweet[3]. Thru twitter everyone either family, friends or officemates can communicate easily and also can stay connected through the exchange of quick and frequent messages[4].

Basically Twitter also is one of the top social networking sites in Malaysia, while normally registered user will use English or Malay language as media of communication. In fact, while use Malay language in twitter also had the same problem with casual written English language as per highlighted by Clark & Araki 2011 which is does not confirm the rules of spelling, grammar, punctuation, misspelled, use the abbreviation language and so on[5]. Meanwhile Samsudin et al. 2012 justify and identified have 13 patterns of common noisy terms in Malay Language[6]. Besides that, Hong, Convertino, & Chi, 2011 identified that Malay language also the higher leading language which used to communicate over the Twitter, Malay language is fourth leading language worldwide in twitter messages broadcasting[3-7]. Table 1 shows same sample of noisy text of Malay tweet.

IOP Conf. Series: Journal of Physics: Conf. Series **806** (2017) 012008 doi:10.1088/1742-6596/806/1/012008

Table 1. Sample noisy token Malay tweet

Original tweet: Akum Slh kew aq inGin cyunx duE OwAnk Yg BERBeZA RUpE?? Aq pown lyAl uNtOk coorUnx! Tutup akum
Malay normalize: Assalamualaikum Salah ke aku ingin sayang dua orang yang berbeza rupa? Aku pun layak untuk kau orang. Waalaikumusalam
Translate: Hi It is wrong, if I fall in love with two difference people? I also eligible for you all. Bye

Based on table 1, can be justify there are several issues that makes the normalization of Malay tweet become a problematic and very difficult task to normalize all the texts. Based on this tweets conversation written extremely colloquially, novel words and interjections[8]. Besides that, a word written using a phonetic spelling ("idop" instead of hidup), and combined with other words into an acronym ("tutup akum" instead of waalaikumussalam)[8]. Other than that, this tweet written with mixed uppercase and lowercase letters. Tweets need to be normalize and replace or remove as much as possible of noisy token or doing spell correction, then convert them to the standard Malay language.

2. Related works

The noisy text normalization methodologies can be written off as into five clusters. The SMT (statistical machine translation) paradigms firstly was familiarized by Aw, Zhang, Xiao and Su in 2006 and also become as the first paradigm's in text normalization[9-10].This paradigm is used to address normalization problems using statistical machine translation task. This approach visualize as a solution for paraphrase problem where a messages in the SMS language will be transformed to standard English language via a similar phrase-based statistical MT method [10]. In other word this paradigm use to normalize SMS text that translates a source language (UGC) to a target language (standard language)[4]. Time by time this technique has been upgraded and enhanced by further researches[11].

The second group of paradigms approaches is Spell-Checker metaphor. The Spell-Checker metaphor focusing on normalize SMS language, which is performed a correction on sequence of words and really concerns about out of vocabulary tokens from 'noisy' input token[12]. This noisy channel model metaphor was introduced by Choudhury, Saraf, Jain, Sarkar, and Basu in 2007, which this method was used a hidden Markov model (HMM)[12]. This method has been used to characterize the stochastic properties of the noisy channel. Further researches from other researchers has been refined and improved this method [13-14]. Besides that, this approach also have been modified by Cook and Stevenson (2009), the modification to design an unsupervised method using probabilistic models[3-15]. In addition, the spell checking approach have been merged with the SMT-like by Beaufort, Roekhaut, Cougnon, and Fairon (2010) for normalize French SMSs[16].

ASR metaphor represent as third group of paradigm approaches, this metaphor has been introduced by Kobus, Yvon, and Damnati 2008 . Kobus and others researcher introduced this metaphor to orthography of French SMS messages. This paradigm was combination of two metaphor which is an ASR-like system and SMT metaphor. Basically main purpose of this ASR metaphor to improve the accuracy of results normalize French SMS[17]. The fourth groups is the dictionary based normalization metaphor, which is very user friendly to used and able to come out fast result[4]. This approach requires the dictionary designed on the basis of the error and not proper English language used, commonly encountered in social media can be divided into several different categories, and in this regard, a multi-faceted approaches are the most effective way to address the problem[5]. It has been proven that using colloquial dictionary that can overcome some of the complex and sophisticated approach[4-5]. However, its performance is very depending on the size of the dictionary. Because of that, method to automatically compile a large data of dictionary has been introduced by Han, Cook, and Baldwin (2012)[18]. While identify the weaknesses in the dictionary approach, Oliva, Serrano, Del Castillo, and Igesias (2013) introduced a special Spanish phonetic dictionary, in which each entry is formed by a coded consonant string, vowels strings, and their positions in the word, for normalizing Spanish SMS texts[4-19].

The fifth group is the Han and Baldwin who described lexical methods for normalize tweet message, this paradigm has been introduced in 2011. For this metaphor, Han and Baldwin implemented combination of four method for normalize the texts, which are; 1). Pre-processing; 2). Confusion set generation; 3). Ill-formed word detection and lastly 4). Candidate selection using a variety of metrics: lexical edit distance, phonemic edit distance, longest common subsequence (LCS), affix substring, language model, and dependency-based frequency features[20]. In 2011 also, Wei, et al enhanced HanBaldwin paradigm's to put an account of time sensitive query on twitter while doing normalization[21]. This encouraged us to plan a recovery methods which has three main phases: pre-processing, generation of candidates, and the selection of candidates[4].

3. Proposal Approach

3.1 Contextual Dictionary

Basically, most of people will using contextual cues to guessing the meaning of unfamiliar words[22]. Based on Knight, 1994, this research justified that normally dictionary will used into two major categories: vocabulary learning and reading comprehension[22]. In lines with the times, nowadays dictionary not only used for translating a language such as from Malay to English, but it also can translate colloquial language. Essentially, dictionary also will represented as database especially while using colloquial dictionary for normalize the texts.

Clark & Araki, 2011 was a pioneer in introduced dictionary metaphor to normalize the text. Their research tried to overcome the disambiguate word problem with the context-aware dictionary[5]. Meanwhile Basri et al. 2012 introduced dictionary normalization system which is contains eight modules in a pipeline for normalize Malay text[23]. After go through this research for every step of modules, can be identified that some of modules not suitable to be execute because will causes of data loosing and inaccurate normalization. Then at the same year Samsudin et al. 2012 ,develop a Malay noisy dictionary. They collecting almost 15000 data from various sources such as tweeter, online forum and Facebook after that manually normalize that words[6].

Basically based on research stated above, either from Clark, Basri or Samsudin and their researcher team, can be justified that this method or metaphor really depending on the size of dictionary database. Because of that, at the same year Han, et al. 2012 come out new method of research which are to extract the OOV and IV pairs based on distributional similarity from English Tweets then after that they select best pairs by string and contextual similarities of words[24]. The longer OOV words in sequence, this technique works more effective. Meanwhile based on Oliva, et al. 2013, this method proposed exercise of the lexical and semantic disambiguation using a combination of lexical resources such as SMS and Spanish dictionaries or the lexical database WordNet[19].

3.2 Decision Making via Language Model

Language model is one of the techniques in determining and compute the possibility of a sentence or sequences of word[24-27]. Originally language model was developed for overcome speech recognition problem[26]. After that this technique widely used in many areas also such as spelling correction, handwriting recognition, machine translation and optical character recognition[25]. A statistical language model be able to represent by the conditional probability of the next words based on sequence words. In the development of statistical model of natural language, the difficulty of this modelling problem is greatly reduced by taking advantage of word order, and the wordy fact that chronology of closer words are statistically dependant[28]. Therefore, n-gram models assemble tables of conditional prospect of the next word, for each one contexts large number[28].

Generally N-grams can be used in a multiplicity of different tasks. While developing the language model, normally n-gram is used to develop unigram model and it's not limited to use on that model only because it also can be used to develop bigram and trigram models. As example, Google and

Microsoft already urbanized n-gram models to the scale web, which are it can be used in a variation of tasks such as spelling correction, word-breaking and text identification. Besides that, from Gu & Berleant 2000 research, they identified that n-gram model also have been used as yearly as 1979 in so many task such as for language identification, spelling correction, document categorization, document comparison, robust handling of noisy (misspelled, OCR'ed etc.) texts, topic highlighting , document space visualization, spoken document retrieval, and other information retrieval related applications[29].

While doing the decision making via language model, feature selection must take an account also because feature selection is the process of selecting a subset of the conditions prevailing in the training set and only use this as a subset in text classification[30]. Feature selection has two main objectives. First, it makes training and use of more efficient classifiers to reduce the effective size of the vocabulary[30]. Second, the selection often improves the classification accuracy by eliminating noise characteristics[30]. Further explanation will elaborate at architecture section.

3.3 Architecture

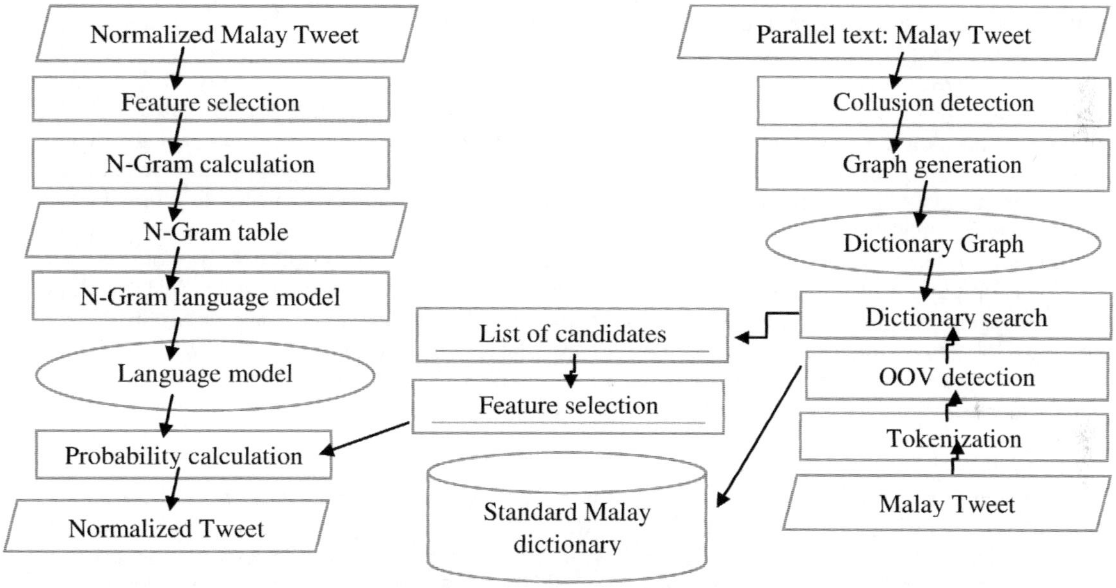

Figure 1. Proposed architecture of Statistical Dictionary Malay tweet normalization

Figure 1 represents proposed architecture of Hybrid Dictionary/Modelling Malay tweet normalization. The first important thing when doing normalization is to tokenize the text. As per suggested by Saloot, Idris & Mahmud 2014, in normalize Malay Twitter message are need to convert all capital letter in setences into small letter, it because capital letter deos not have orthographic value[4]. This architecture also will implementing tokenization module has been introduced by Saloot, Idris & Mahmud 2014. Based on this architecture the second step will take an action is to descriminate OOV words and IV words. While detecting OOV words, we will checking the words by comparing with standard Malay dictionary database. In this OOV detection module will also implementing a set of XXX heuristic rules which has been introduced by Saloot, Idris & Aw 2014[3]. Besides that, C ++ vector data structure will be implementing to speed up the process and combined module responsible for generating candidate, modified lexical and phonemically[3]. This module is detected OOV sent word to the lexical components and applying a phoneme module to module lexical decision[3].

Actually when entering Malay tweet into this architecture, there are two jobs will process in parallel concurrently in order to get final result of normalized Malay tweet;1). Parallel text:Malay

Tweet; 2). Normalized Malay tweet. From the parallel text: Malay tweet proses flow sequences, first procedure is entering the Malay twitter message pair together with standard Malay language. Collusion detection method will used because it practice of two or more parties to work together where they have intentionally harm others[31]. While doing collusion detection, standard Malay language can be a benchmark for generating the graph, some algorithm will be create for graph generation[32]. From the study, we identified that Kronecker product of two matrices was the perfect instrument for achieve parallel text normalization goal[33]. After that dictionary graph will be created, the purpose of generate dictionary graph is for dictionary search, where it will work together with OOV detection, which is list of Malay words with the common proper name of cities and countries as example[3]. From the dictionary search, the list of candidates or result of the correct words will come out and again feature selection will be practices.

Meanwhile, at the same time in parallels this architecture also will be processing the second jobs which is normalization Malay tweet. Here, first feature selection will be implementing to making n-gram calculation. From the research Gu & Berleant 2000 justified because of uniqueness of strings of length n, n-gram can reduce the match word problem, also able to enhanced computational effectiveness compared with the word processing[34]. For the reason of the strings of n length, it can support until 5- gram even more and can be so challenging, that way hash array n-gram table need be create to store large amounts of texts[25-29]. Besides that, N-gram support partial matches if text contains words that are different but similar words because of the similarity between the words sequence having n-grams in common[29]. Creating N-gram tables was one of the step for making N-gram language modelling, because model whose assigning probabilities to sequence of words named as Language Model[28]. Actually what n-gram language model will do is to breaks up the sentence into smaller sequences of words and figures up the individual n-gram probabilities. The order of the words very important in Language Model and it can combine in order to create the LM by making weaker independence assumptions[34]. Zhu 2007, justified that the acclimatizing part $wi-n+1:i-1$ is known as 'history', which $n - 1$ was represent previous words[34]. Basic formula for probability calculation n-gram language model. After make a probability calculation based on list of candidates from feature selection result, we can get final result of normalized texts.

4. Discussion

This research paper only presenting the conceptual level of architecture because it still work-in-progress research. The aim of this research is to develop a hybrid dictionary/modelling approach for Malay Tweet Normalization. Meanwhile this research have three main objectives: - 1). To propose new architecture based on statistical approach for normalizing Malay tweet. 2). To develop a prototype of hybrid dictionary model to normalize Malay tweet. 3). To evaluate the result/system (prototype) with BLUE evaluation technique and also validate the result of normalization Malay tweet. The architecture of this research will be contributed by my research on Malay Text Normalization.

5. Future work and Conclusion

This paper proposed a hybrid modelling/dictionary approach for Malay Tweet normalization. Based on purposed architecture/ modelling, this research will implementing machine learning which "gives computers the ability to learn without being explicitly programmed"[35]. This research target to handle all the problem or tasks while normalizing Malay Tweet likes supervised learning, unsupervised learning and reinforcement learning using hybrid dictionary and decision making via language model and n-grams model. For the future works, more detail and further experimental result elaborated in the next extended version of this paper.

References

[1] Dasselaar, C., 2016. English Versus Native Language in Digital Diplomacy.
[2] A. Mislove, S. Lehmann, Y.-Y. Ahn, J.- P. Onnela, and J. N. Rosenquist, "Understanding the Demographics of Twitter Users.," in ICWSM, 2011.
[3] Saloot, M.A., Idris, N. & Aw, A., 2014. Noisy Text Normalization Using an Enhanced Language Model. In Proceedings of the International Conference on Artificial Intelligence and Pattern Recognition. Kuala Lumpur, Malaysia: SDIWC, pp. 111–122.
[4] Saloot, M.A., Idris, N. & Mahmud, R., 2014. An architecture for Malay Tweet normalization. Information Processing and Management, 50(5), pp.621–633.
[5] Clark, E. & Araki, K., 2011. Text normalization in social media : progress , problems and applications for a pre-processing system of casual English. , 27(Pacling), pp.2–11.
[6] Samsudin, N. et al., 2012. Normalization of Common NoisyTerms in Malaysian Online Media. Proceedings of the Knowledge Management International Conference, (July), pp.515–520.
[7] Hong, L., Convertino, G., & Chi, E. H. (2011). Language matters in Twitter: A large scale study. In L. A. Adamic, R. A. Baeza-Yates, & S. Counts (Eds.), ICWSM. The AAAI Press.
[8] Kaufmann, M., 2010. Syntactic Normalization of Twitter Messages. , pp.1–7.
[9] Aw, A., Zhang, M., Xiao, J., & Su, J. (2006). A phrase-based statistical model for SMS text normalization. In Proceedings of the COLING/ACL on main conference poster sessions (pp. 33–40). Stroudsburg, PA, USA: Association for Computational Linguistics.
[10] Saloot, Idris, Shuib, R.,(2015) Towards Tweets Normalization Using Maximum Entropy.
[11] Lopez Ludeña, V., San Segundo, R., Montero, J. M., Barra Chicote, R., & Lorenzo, J. (2012). Architecture for text normalization using statistical machine translation techniques. In IberSPEECH 2012 (pp. 112–122). Madrid, Spain:
[12] Kobus, C., Yvon, F. & Damnati, G., 2008. Normalizing SMS: Are Two Metaphors Better Than One? In Proceedings of the 22Nd International Conference on Computational Linguistics - Volume 1. COLING '08. Stroudsburg, PA, USA: Association for Computational Linguistics, pp. 441–448.
[13] Choudhury, M., Saraf, R., Jain, V., Mukherjee, A., Sarkar, S., & Basu, A. (2007). Investigation and modeling of the structure of texting language. International Journal of Document Analysis and Recognition (IJDAR), 10(3), 157–174. http://dx.doi.org/10.1007/s10032-007-0054-0.
[14]Liu, F., Liu, Y. & Weng, F., 2011. Why is "SXSW " trending ? Exploring Multiple Text Sources for Twitter Topic Summarization. Proceedings of the Workshop on Language in Social Media, (June), pp.66–75.
[15]Xue, Z., Yin, D. & Davison, B., 2011. Normalizing Microtext. Analyzing Microtext, pp.74–79.
[16]Cook, P. & Stevenson, S., 2009. An Unsupervised Model for Text Message Normalization. , (June), pp.71–78.
[17]Beaufort, R. et al., 2010. A hybrid rule / model-based finite-state framework for normalizing SMS messages. , 1(July), pp.770–779.
[18]Han, B., Cook, P. & Baldwin, T., 2012. Automatically constructing a normalisation dictionary for microblogs. EMNLP-CoNLL 2012 - 2012 Joint Conference on Empirical Methods in Natural Language Processing and Computational Natural Language Learning, Proceedings of the Conference, (July), pp.421–432.
[19]Oliva, J. et al., 2013. A SMS Normalization System Integrating Multiple Grammatical Resources. Natural Language Engineering, 19(1), pp.121–141.
[20]Han, B. & Baldwin, T., 2011. Lexical Normalisation of Short Text Messages : Makn Sens a # twitter. , pp.368–378.
[21]Z. Wei, L. Zhou, B. Li, K.-F. Wong, and W. Gao, "Exploring Tweets Normalization and Query Time Sensitivity for Twitter Search.," in TREC, 2011.
[22]Knight, S., 1994. Dictionary Use While Reading : The Effects On Comprehension and Vocabulary Acquisition For Students Of Different Verbal Abilities. , (47).

[23]Basri, S.B., Alfred, R. & On, C.K., 2012. Automatic spell checker for Malay blog. In Control System, Computing and Engineering (ICCSCE), 2012 IEEE International Conference on. pp. 506–510.
[24]Collins, M., 2013. Language Modeling.
[25]Goodman, J., 2001. A Bit of Progress in Language Modeling. Technical Report, p.73.
[26]Watanabe, S. et al., 2011. Topic tracking language model for speech recognition. Computer Speech & Language, 25(2), pp.440–461.
[27]Jurafsky, D., Language Modeling. , p.88.
[28]Bengio, Y. et al., 2003. A Neural Probabilistic Language Model. The Journal of Machine Learning Research, 3, pp.1137–1155.
[29]Gu, Z. & Berleant, D., 2000. Technical Report 10-00a, Oct. 2000, Software Research Lab, 3215 Coover Hall, Dept. of Electrical and Computer Engineering, Iowa State University, Ames, Iowa 50011 USA. , pp.1–12.
[30]Manning, C.D. & Raghavan, P., 2009. An Introduction to Information Retrieval A. C.-B. E. Salas, ed. Online.
[31]Mazrooei, P., 2013. Collusion detection in sequential games. Proceedings of the Twenty-Seventh AAAI Conference on Artificial Intelligence, MR90652, p.88
[32]Cordeiro, D. et al., 2010. Random graph generation for scheduling simulations. Proceedings of the 3rd International ICST Conference on Simulation Tools and Techniques, p.60.
[33]Leskovec, J., Chakrabarti, D. & Kleinberg, J., Realistic , Mathematically Tractable Graph Generation and Evolution , Using Kronecker.
[34]Zhu, X., 2007. CS838-1 Advanced NLP : Language Modeling The Need for Finding Likely Sentences. Language, (3), pp.1–7.
[35]"Machine Learning: What it is and why it matters". *www.sas.com*. Retrieved 2016-12-29.

IOP Conf. Series: Journal of Physics: Conf. Series **806** (2017) 012009 doi:10.1088/1742-6596/806/1/012009

Generation of Test Questions from RDF Files Using PYTHON and SPARQL

Assel Omarbekova, Altynbek Sharipbay and Alibek Barlybaev

Faculty Of Information Technologies, L.N Gumilyov Eurasian National University, Astana, Kazakhstan

omarbekova@mail.ru
sharalt@mail.ru
frank-ab@mail.ru

Abstract. This article describes the development of the system for the automatic generation of test questions based on the knowledge base. This work has an applicable nature and provides detailed examples of the development of ontology and implementation the SPARQL queries in RDF-documents. Also it describes implementation of the program generating questions in the Python programming language including the necessary libraries while working with RDF-files.

1. Knowledge base development

The aim of the State Program of Development of Education of the Republic of Kazakhstan for 2011-2020 is to increase the competitiveness of education, human capital development by providing quality education for sustainable economic growth. One of the main directions of the program is the electronic learning «e-learning». The implementation of the State Program of Education Development Program of Kazakhstan for 2011-2020 formulated the following task: for each subject studied in primary and specialized school, interactive and intellectual digital educational resources will be developed.

Therefore, nowadays there is a need of the application of artificial intelligence technology in development of educational resources and in the learning process. Currently, the Research Institute "Artificial Intelligence" of L.N Gumilyov Eurasian National University conducts research in this area, developing a knowledge base on the various subject areas. Let's look at the possibility of automatic generation of test questions from the knowledge base.

RDF is developed for computers, so they could understand and read it. It isn't for people and display RDF description on the Internet. RDF is written on XML (Extended Mark-up Language), and it's also a semantic web activity of W3C.

The ontology «ComputerScience.rdf» was developed. RDFS/RDF classes, also the options, like label, domain, range and comment, and attributes, such as about, description, resource datatype, are used in this document [1].

After designing RDF-document «ComputerScience.rdf», the file should be checked for validity using a web service «RDF Validator» on the web-site «www.w3.org». The system displays a list of triples and graph data model. A detail of the generated graph after successful validation of the document «ComputerScience.rdf» is shown in Figure 1. We can see the subjects «http://learningsparql.com/ns/myExample#Database» and «http://learningsparql.com/ns /myExample#Algorithm», eight predicates and eight objects related to their subjects.

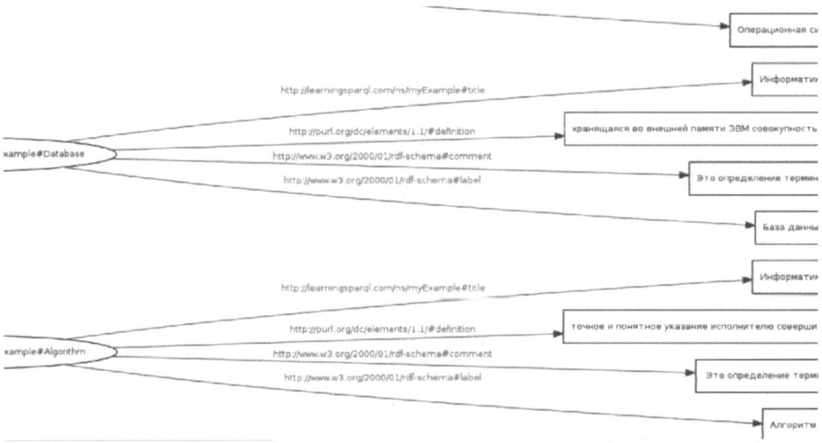

Figure 1. Enlarged part of graph «ComputerScience.rdf»

2. Generation of test questions for the evaluation of students' low-lever skills

The development of test questions for the learning progress evaluation of students requires a serious approach. Tests of high quality are designed carefully so that they are not ambiguous or obscure for the test takers [2]. In the development of test tasks we have investigated various techniques and also deeply studied Bloom's taxonomy. According to the taxonomy, the cognitive skills of pupils are divided into six categories, such as knowledge, comprehension, application, analysis, synthesis and evaluation. The most primitive of the skills are knowledge and comprehension, and the most advanced skills are analysis, synthesis and evaluation. In the development of test tasks, it was decided to generate test questions to test low-level skills in the knowledge and comprehension. Because generating questions for analysis and evaluation a system is needed that is able to make arguments in full. According to Bloom's taxonomy special verbs are defined for the development of test tasks. Some verbs for knowledge and determination, like 'give the definition', 'list', 'name', 'explain', etc. With the use of these verbs we can estimate the low-level skills, while verbs, like 'classify', 'locate' and 'describe the advantages' are used to evaluate higher level skills of students.

Program realization starts with installment of IDLE Python 3.5 and «rdflib» library, to develop graphic interface the library «tkinter» is imported into the code. The code of generating test questions of evaluation of low-level skills is shown below.

```
from tkinter import *
import tkinter.messagebox
import rdflib
root = Tk()
def generate():    g = rdflib.Graph()
   g.parse(file_path)
   qres = g.query("""
      PREFIX rdfs: <http://www.w3.org/2000/01/rdf-schema#>
      PREFIX ns0: <http://learningsparql.com/ns/addressbook#>
      SELECT ?t ?def
      WHERE   {?a dc:title ?t;        dc:definition ?def. }""")
   global item_arr
   item_arr = []
   for row in qres:        items ="Define the term ' "+str(row[0])+"'"
      item_arr.append(items)
   for item in range(len(item_arr)):
      str_number = str(item+1)
      the_item = str_number+". "+item_arr[item]
```

```
        global label
        label = Label(frame, text = the_item, width="200",height="2", bg="grey", fg="white", font=("Helvetica", 11), justify=LEFT, anchor=W)
        label.pack()
    def openFile():    root.fileName = filedialog.askopenfilename(filetypes=(("RDF files", ".rdf"), ("OWL files", ".owl"), ("All files", "*.*")))
        global file_path
        file_path = root.fileName
        return file_path
    def quit_program():   exit()
    def show_info():    tkinter.messagebox.showinfo('Window title', 'This application generates low order thinking skills questions')
    root.geometry("600x400+300+200")
    root.title("Automated test items generator")
    frame = Frame(root, bg="sky blue", )
    frame.pack()
    menu = Menu(root)
    root.config(menu=menu)
    subMenu = Menu(menu)
    menu.add_cascade(label="File", menu=subMenu)
    subMenu.add_command(label="Open file", command=openFile)
    subMenu.add_separator()
    subMenu.add_command(label="Exit", command=quit_program)
    helpMenu = Menu(menu)
    menu.add_cascade(label="Help", menu=helpMenu)
    helpMenu.add_command(label="Info", command=show_info)
    button1 = Button(frame, text="GENERATE", fg="blue", command=generate)
    button1.pack(side="top", fill='both', expand=True, padx=4, pady=4)
    root.mainloop()
```

The system allows to open any file of rdf and owl type. Choose the file ComputerScience.rdf and click "Generate". The questions based on a knowledge base will be shown. The form is on Figure 2.

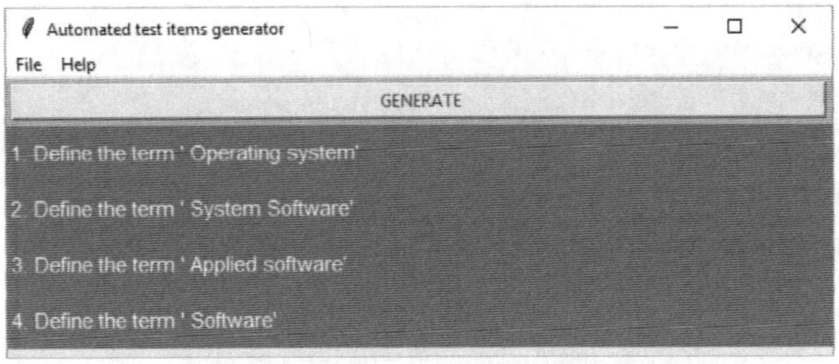

Figure 2. A form with questions testing the knowledge of terms.

3. Generation of test questions to evaluate high-level skills of students

After developing the test questions to evaluate low-level skills, the small owl file under flowers_ontology.owl was developed [3]. A fragment can be seen on Figure 3.

IOP Conf. Series: Journal of Physics: Conf. Series **806** (2017) 012009 doi:10.1088/1742-6596/806/1/012009

```
<owl:Class rdf:about="http://www.linkeddatatools.com/plants#shrubs">

    <!-- Shrubs is a subclassification of planttype -->

    <rdfs:subClassOf rdf:resource="http://www.linkeddatatools.com/plants#planttype"/>
    <rdfs:label>Shrubbery</rdfs:label>
    <rdfs:comment>Shrubs, a type of plant which branches from the base.</rdfs:comment>

</owl:Class>

<!-- Individual (Instance) Example RDF Statement -->
<rdf:Description rdf:about="http://www.linkeddatatools.com/plants#magnolia">

    <!-- Magnolia is a type (instance) of the flowers classification -->
    <rdf:type rdf:resource="http://www.linkeddatatools.com/plants#flowers"/>
    <rdfs:label>Magnolia</rdfs:label>
    <rdfs:comment>Magnolia, a type of flower.</rdfs:comment>

</rdf:Description>

<rdf:Description rdf:about="http://www.linkeddatatools.com/plants#rose">

    <!-- Rose is a type (instance) of the flowers classification -->
    <rdf:type rdf:resource="http://www.linkeddatatools.com/plants#flowers"/>
    <rdfs:label>Rose</rdfs:label>
    <rdfs:comment>Rose, a type of flower.</rdfs:comment>

</rdf:Description>

<rdf:Description rdf:about="http://www.linkeddatatools.com/plants#aloe">
```

Figure 3. A file fragment flowers_ontology.owl

Below, on the Figure 4, we can see a generated graph's fragment from the file flowers_ontology.owl

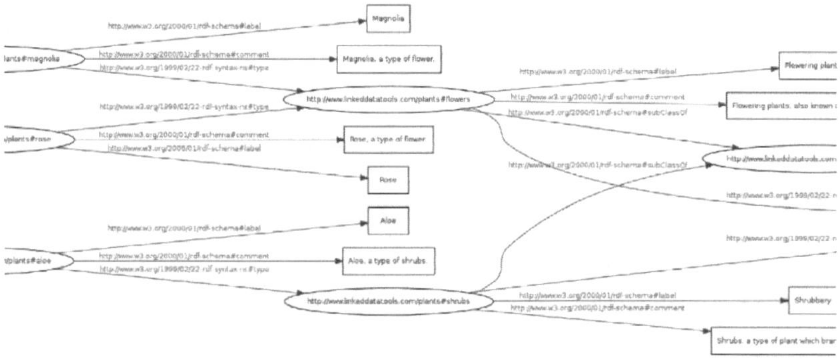

Figure 4. A fragment of generated graph from the file flowers_ontology.owl

The following code allows to generate questions to classify objects.

```
from tkinter import *
import tkinter.messagebox
import rdflib
root = Tk()
def generate():   g = rdflib.Graph()
   g.parse(file_path)# путь к файлу
   qres = g.query("""
   PREFIX   rdfs: <http://www.w3.org/2000/01/rdf-schema#>
   PREFIX    rdf: <http://www.w3.org/1999/02/22-rdf-syntax-ns#>
   PREFIX plants: <http://www.linkeddatatools.com/plants#>
   PREFIX     dc: <http://purl.org/dc/elements/1.1/#>
   PREFIX    owl: <http://www.w3.org/2002/07/owl#>
    SELECT ?name
   WHERE
{?entity rdf:type/rdfs:subClassOf* plants:planttype;
      rdfs:label ?name.} LIMIT 5""")
   # cycle to text output  on console
   global item_arr
```

```
    item_arr = []
    for row in qres:        items = str(row[0])
      item_arr.append(items)
    global label
  label = Label(frame, text = " Classify the following terms. To what class they belong?", width="200",height="2", bg="blue", fg="white", font=("Helvetica", 11), justify=LEFT, anchor=W)
    label.pack()
    for item in range(len(item_arr)):        str_number = str(item+1)
      the_item = str_number+". "+item_arr[item]
      label = Label(frame, text = the_item, width="200",height="2", bg="grey", fg="white", font=("Helvetica", 11), justify=LEFT, anchor=W)
      label.pack()
  def openFile():      root.fileName = filedialog.askopenfilename(filetypes=(("RDF files", ".rdf"), ("OWL files", ".owl"), ("All files", "*.*")))
    global file_path
    file_path = root.fileName
    return file_path
  def quit_program():    exit()
  def show_info():    tkinter.messagebox.showinfo('Window title', 'This application generates low order thinking skills questions')
  root.geometry("600x400+300+200")
  root.title("Automated test items generator")
  frame = Frame(root, bg="sky blue", )
  frame.pack()
  menu = Menu(root)
  root.config(menu=menu)
  subMenu = Menu(menu)
  menu.add_cascade(label="File", menu=subMenu)
  subMenu.add_command(label="Open file", command=openFile)
  subMenu.add_separator()
  subMenu.add_command(label="Exit", command=quit_program)
  helpMenu = Menu(menu)
  menu.add_cascade(label="Help", menu=helpMenu)
  helpMenu.add_command(label="Info", command=show_info)
  button1 = Button(frame, text="GENERATE", fg="blue", command=generate)
  button1.pack(side="top", fill='both', expand=True, padx=4, pady=4)
  root.mainloop()
```

Figure 5 illustates a program with graphic interface generating test tasks to classify plants.

After the development of the system, as a result of the testing an error has not been found, and the system regularly generated test items from the knowledge base on the various subject areas. When testing the system several different RDF files were used, in various formats such as, of OWL and TTL. According to test results, rdflib library for the Python programming language, connected to process RDF files, it was not able to work with TTL files. Nevertheless, the system processed RDF and OWL files using the SPARQL query language for RDF documents [4,5].

Also, tests were run and it was attempted to create test questions generated on the basis of the arguments using owlready libraries for python programming language. This library includes Hermit reasoning engine that allows drawing conclusions in respect of certain resources to specific classes. As a result of attempts to generate questions based on logical reasoning the desired results weren't recieved. Request made to obtain the logical conclusion was made on the owl files, but they did not give the desired results.

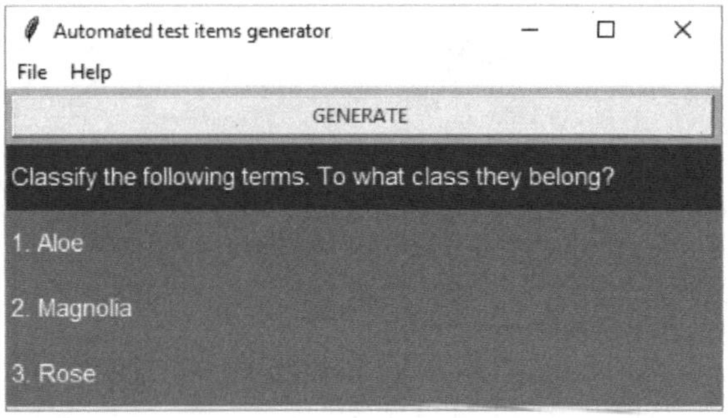

Figure 5. Graphic interface of program with a question

4. Conclusion

Summing up we can draw conclusions regarding the relevance and trends in the development of this direction. Today, all over the world there are a lot of work in the direction of artificial intelligence and intelligent systems based on knowledge [6]. The research shown in this paper provides new ideas and opens up new possibilities of intelligent systems based on knowledge bases for further work and study.

As a result of research the following tasks were solved:

- Development of a knowledge base in RDF format.
- Development of software applications using the IDLE Python to generate test questions.
- Creation of the SPARQL query through IDLE Python in the knowledge base in RDF format.
- Use extracted data from the knowledge base for automatic generation of test questions to test low-level and high-level skills on student knowledge and understanding of the subject area.

The development of systems capable of making inferences from the knowledge base is developing dynamically worldwide. Removing the existing knowledge and generating new knowledge and approving of the existing knowledge base are made by means of specialized engines designed to discuss and draw logical conclusions. The main idea of this work was to extract knowledge from the knowledge base using various latest development tools based on knowledge systems. The system for generating test questions from the knowledge base is one of the steps towards the automation of the evaluation process.

References

[1] DuCharme B 2013 *Learning SPARQL, Querying and Updating with SPARQL* (O`Reilly) pp 3 - 200
[2] Julita B A 2007 *Simplified Guide to Create an Ontology* (ASLab R-2007-004 vol 1)
[3] Negnevitsky M 2002 *Artificial Intelligence – A Guide to Intelligent Systems* (Addison-Wesley Publishing Company. Edinburgh)
[4] William J T 2010 *Artificial Intelligence – Agent Behaviour* p 138
[5] Corby O and Faron-Zucker C 2007 *RDF/SPARQL design pattern for contextual metadata* (Proceedings of the IEEE/WIC/ACM international conference on web intelligence) pp 474-477
[6] Leida M and Chu A 2013 *Distributed SPARQL query answering over RDF data streams* (IEEE international congress on BIG DATA) pp 369-378

An Improved Differential Evolution Algorithm and Its Application to Large-Scale Artificial Neural Networks

Tae Jong Choi and Chang Wook Ahn

Department of Electrical and Computer Engineering, Sungkyunkwan University, 2066, Seobu-ro, Jangan-gu, Suwon-si, Gyeonggi-Do, Republic of Korea

cwan@skku.edu

Abstract. A new differential evolution (DE) algorithm is presented in this paper. The proposed algorithm monitors the evolutionary progress of each individual and assigns appropriate control parameters depends on whether the individual is successfully evolved or not. We conducted the performance evaluation on CEC 2014 benchmark problems and confirmed that the proposed algorithm outperformed than the conventional DE algorithm. In addition, we apply the proposed DE algorithm as an optimization technique of training large scale multilayer perceptron. We conducted the performance evaluation on an artificial neural network that has approximately 1,000 weights and confirmed again that the proposed algorithm performed better than the conventional DE algorithm. As a result, we proposed a new DE algorithm that has better optimization performance for solving large-scale global optimization problems.

1. Introduction

Training artificial neural networks[1,2] is considered as a multimodal continuous optimization problem. Generally, Back-propagation algorithm is used for this purpose. However, this algorithm has a disadvantage that it has the higher probability to get stuck in local minima. Some researchers attempt to devise a method to overcome this problem by using another optimization technique such as evolutionary algorithms[3-6].

In this paper, we propose a new differential evolution (DE) algorithm and apply this algorithm to training large-scale artificial neural networks as an optimization technique. DE algorithm has three control parameters and these control parameters affect its optimization performance significantly. However, it is the laborious task to apply the trial-and-error approach to find appropriate control parameters depends on the structure of artificial neural networks.

Therefore, the proposed DE algorithm automatically finds and tunes appropriate control parameters. More specifically, the proposed algorithm monitors the evolutionary progress of each individual. If an individual is successfully evolved to the next generation, then the individual is assigned new control parameters tuned by the Gaussian distribution. If an individual is failed to evolve, then the individual is assigned new control parameters tuned by the Cauchy distribution.

We conducted the performance evaluation on CEC 2014 benchmark problems and confirmed that the proposed algorithm outperformed than the conventional DE algorithm. In addition, we apply the proposed DE algorithm as an optimization technique of training large scale multilayer perceptron. We conducted the performance evaluation on an artificial neural network that has approximately 1,000 weights and confirmed again that the proposed algorithm performed better than the conventional DE algorithm.

2. Background

2.1. Artificial Neural Networks

Generally, Back-propagation algorithm[1,2] is used for training artificial neural networks. First, the output value generated from the input layer, the hidden layer, and the output layer is compared with the target value. And adjust the weights from the output layer to the hidden layer through the differential by the difference that occurs. We then adjust the weights by repeating this process up to the input layer. In this process, the chain-rule of the differential is used. Back-propagation algorithm, however, has disadvantages. A typical example is that there is a high probability of convergence to local optima and cannot escape from the saddle-point.

There are studies that use evolutionary algorithms as a way to overcome these shortcomings. Evolutionary algorithms do not need to consider differentiability and the above mentioned disadvantages disappear. For this reason, some researchers have been studying artificial neural network training methods using evolutionary algorithms. However, training methods of artificial neural networks using evolutionary algorithms also have disadvantages. In particular, it is considered that evolutionary algorithms are not suitable for learning artificial neural networks with large-scale weights. The reason for this is that it requires a lot of computational cost to apply vast amounts of data to existing evolutionary algorithms.

Recent study has confirmed that evolutionary algorithms can perform well similar to Back-propagation algorithm through hypothesis that each individual need not use all training data. In other words, unlike previous methods in which individuals learn all training data, this condition can be mitigated by reducing the computational cost and achieving the same level of performance as Back-propagation algorithm.

2.2. Differential Evolution

DE algorithm [7,8,9] contains NP individuals. $\overrightarrow{X_i^G} = \{x_{i,1}^G, x_{i,2}^G, \dots, x_{i,D}^G\}$ denotes a target vector of D components in G generation. Target vector acts as a parent individual in evolutionary algorithms. $\overrightarrow{U_i^G} = \{u_{i,1}^G, u_{i,2}^G, \dots, u_{i,D}^G\}$ denotes a trial vector. Trial vector acts as a child individual. And, $\overrightarrow{V_i^G} = \{v_{i,1}^G, v_{i,2}^G, \dots, v_{i,D}^G\}$ denotes a mutant vector. Mutant vectors are used to generate trial vectors with target vectors.

The DE algorithm consists of mutation, crossover, and selection operators. The mutant operator generates a mutant vector based on three donor vectors selected randomly in the population. This is as follows.

$$\overrightarrow{V_i^G} = \overrightarrow{X_{r_1}^G} + F \cdot \left(\overrightarrow{X_{r_2}^G} - \overrightarrow{X_{r_3}^G}\right) \quad (1)$$

where $r_1, r_2, r_3 \in \{1,2,\dots,NP\}$, $r_1 \neq r_2 \neq r_3 \neq i$. F is scaling factor, which is one of the control parameters of DE algorithm. In the crossover operator, a trial vector is generated based on a target vector and a mutant vector. This is as follows.

$$u_{i,j}^G = \begin{cases} v_{i,j}^G, & if\ rand_j[0,1) \leq CR\ or\ j \equiv jrand_i \\ x_{i,j}^G, & otherwise \end{cases} \quad (2)$$

where $jrand_i \in \{1,2,\dots,D\}$. $jrand_i$ ensures that at least one of the components of the trial vector is composed of the components of the mutant vector. CR is crossover rate, which is one of the control parameters of DE algorithm. Selection operator selectes the individuals to be populated in the $G+1$ generation population This is as follows.

$$\overrightarrow{X_i^{G+1}} = \begin{cases} \overrightarrow{U_i^G}, & if\ f\left(\overrightarrow{U_i^G}\right) \le f\left(\overrightarrow{X_i^G}\right) \\ \overrightarrow{X_i^G}, & otherwise \end{cases} \quad (3)$$

where f is an objective function. The DE algorithm executes these operators until the termination condition is satisfied.

3. Improved Differential Evolution

In DE algorithm, the more likely an individual has the appropriate control parameters, the more likely the individual is to evolve successfully. In other words, successfully evolved individuals are likely to have the appropriate control parameters. Using this observation, we propose a new adaptive parameter control method. The proposed method monitors the selection operator that which individuals succeeded in evolution and which individuals failed. The successfully evolved individuals tunes their control parameters using the Gaussian distribution, a short tailed distribution, and the failed individuals tunes their control parameters using the Cauchy distribution, a long tailed distribution. Through this process, the following effects can be obtained. First, even if an individual has been successfully evolved, it searches for a new value in the neighbor of its control parameters and applies it. This helps to find a new value that are better than the current control parameters. Next, an individual that has failed to evolve uses a large step to find and apply a new value that is far from its control parameters. This helps individuals which fail to evolve to find a new control parameter. The adaptive parameter control method of the proposed algorithm is as follows.

$$\overrightarrow{F_i^{G+1}} = \begin{cases} rand_{G_i}\left(\overrightarrow{F_i^G}, 0.1\right), & if\ \overrightarrow{X_i^{G+1}} \ne \overrightarrow{X_i^G} \\ rand_{C_i}\left(\overrightarrow{F_i^G}, 0.1\right), & otherwise \end{cases} \quad (4)$$

$$\overrightarrow{CR_i^{G+1}} = \begin{cases} rand_{G_i}\left(\overrightarrow{CR_i^G}, 0.1\right), & if\ \overrightarrow{X_i^{G+1}} \ne \overrightarrow{X_i^G} \\ rand_{C_i}\left(\overrightarrow{CR_i^G}, 0.1\right), & otherwise \end{cases} \quad (5)$$

where $rand_G(\mu, \sigma^2)$ and $rand_C(x_0, \gamma)$ denote the Gaussian distribution and the Cauchy distribution, respectively. Then, truncate the control parameters as follows. If $\overrightarrow{F_i^{G+1}}$ is lower than 0.1, $\overrightarrow{F_i^{G+1}} = 0.1$, else if $\overrightarrow{F_i^{G+1}}$ is bigger than 1.0, $\overrightarrow{F_i^{G+1}} = 1.0$. If $\overrightarrow{CR_i^{G+1}}$ is lower than 0, $\overrightarrow{CR_i^{G+1}} = 0$, if $\overrightarrow{CR_i^{G+1}}$ is bigger than 1.0, $\overrightarrow{CR_i^{G+1}} = 1.0$. Due to page limitation, we omit the analysis of the proposed algorithm.

4. Experiment and Result

4.1. CEC2014 Benchmark Problems

We performed the performance evaluation of the proposed algorithm in the CEC 2014 benchmark problems [10] with the conventional DE algorithm. Here, the control parameters of the conventional DE algorithm are: $F = 0.5$, $CR = 0.5$. A total of 50 performance evaluations were performed independently. We used the following settings.

- Population Size: $NP = 100$
- Function Evaluation Counter: $10000 \times D \times NP = 3.0E7$

Table 1 shows the experimental results. The algorithms that perform better in this table are shown in bold. As we can see in this table, we can confirm that the proposed algorithm has better optimization performance than the conventional DE algorithm. The proposed algorithm outperformed or exceeded the conventional DE algorithm with the exception of 5 problems for a total of 30 problems. Therefore, it is experimentally confirmed that the adaptive parameter control of the proposed algorithm can improve the performance of the DE algorithm.

Table 1. The result of performance evaluation in CEC2014 benchmark problems, $D = 30$

	Func.	F_1	F_2	F_3	F_4	F_5	F_6	F_7	F_8	F_9	F_{10}
IDE	Avg.	**2.21E+05**	**0.00E+00**	**0.00E+00**	**4.48E+01**	**2.04E+01**	**5.44E+02**	1.28E-03	**0.00E+00**	**3.70E+01**	**8.96E-01**
	Std. Dev.	**1.69E+05**	**0.00E+00**	**0.00E+00**	**3.64E+01**	**5.66E-02**	**2.13E+00**	3.28E-03	**0.00E+00**	**1.56E+01**	**1.20E+00**
DE	Avg.	9.42E+07	1.58E-09	6.71E-05	7.07E+01	2.09E+01	5.71E+02	**3.19E-14**	9.27E+01	1.78E+02	2.74E+03
	Std. Dev.	1.91E+07	6.80E-10	2.64E 05	5.06E-01	5.12E-02	1.60E+00	**5.12E-14**	7.85E+00	9.26E+00	2.00E+02
	Func.	F_{11}	F_{12}	F_{13}	F_{14}	F_{15}	F_{16}	F_{17}	F_{18}	F_{19}	F_{20}
IDE	Avg.	**3.66E+03**	**6.58E-01**	**2.77E-01**	3.13E-01	**6.89E+00**	**1.09E+01**	**3.16E+03**	**6.85E+01**	**4.71E+00**	**2.48E+01**
	Std. Dev.	**2.64E+02**	**8.91E-02**	**4.93E-02**	7.84E-02	**1.12E+00**	**3.00E-01**	**3.08E+03**	**2.49E+02**	**1.05E+00**	**3.70E+01**
DE	Avg.	6.60E+03	2.06E+00	3.92E-01	**2.81E-01**	1.62E+01	1.27E+01	2.30E+06	5.22E+03	7.85E+00	1.64E+02
	Std. Dev.	1.91E+02	2.43E-01	4.52E-02	**4.75E-02**	9.51E-01	1.43E-01	7.68E+05	3.04E+03	1.09E+00	1.95E+01
	Func.	F_{21}	F_{22}	F_{23}	F_{24}	F_{25}	F_{26}	F_{27}	F_{28}	F_{29}	F_{30}
IDE	Avg.	**4.35E+02**	**8.25E+01**	**3.15E+02**	2.28E+02	**2.04E+02**	**1.00E+02**	4.03E+02	8.49E+02	**9.41E+02**	**2.48E+03**
	Std.Dev.	**5.11E+02**	**6.09E+01**	**0.00E+00**	4.56E+00	**8.62E-01**	**0.00E+00**	3.55E+01	3.35E+01	**1.36E+02**	**1.08E+03**
DE	Avg.	1.14E+05	1.72E+02	**3.15E+02**	**2.23E+02**	2.22E+02	**1.00E+02**	**3.35E+02**	**8.29E+02**	2.85E+03	2.86E+03
	Std.Dev.	3.95E+04	5.85E+01	**0.00E+00**	**1.24E+00**	3.52E+00	**0.00E+00**	**3.65E+01**	**2.41E+01**	5.36E+02	6.33E+02

4.2. Training Multilayer Perceptron

We performed the performance evaluation of the proposed algorithm in the training of the multilayer perceptron with the conventional DE algorithm. The dataset we used is CANCER [11]. The CANCER dataset is divided into a total of 30 attributes and two classes. There are 569 instances in CANCER, of which 75% is used as training data and the remaining 25% is used as test data. Here, the control parameters of the conventional DE algorithm are: $F = 0.5$, $CR = 0.5$. A total of 10 performance evaluations were performed independently. We used the following settings.

- Network Structure: [30-20-10-2], [30-10-2]
- Population Size: $NP = 10$ for [30-20-10-2], $NP = 100$ for [30-10-2]
- Function Evaluation Counter: 500 for [30-20-10-2], 1000 for [30-10-2]

In the network structure of [30-20-10-2], each individual has a total of 620 + 210 + 22 = 852 chromosomes including bias. In the network structure of [30-10-2], each individual has a total of 310 + 22 = 332 chromosomes including bias. Therefore, the training multilayer perceptron problem is a large scale optimization problem. Table 2 shows the experimental results. The algorithms that perform better in this table are shown in bold. As we can see in this table, we can confirm that the proposed algorithm has superior classification performance compared to the conventional DE algorithm.

Table 2. The result of performance evaluation in the training of the multilayer perceptron

	Algorithm	Network Structure	Population Size	Result	
				Avg.	Std.Dev.
CANCER	IDE	30-20-10-2	10	**3.16.E-01**	**4.61.E-04**
	DE	30-20-10-2	10	5.00.E-01	1.41.E-01
CANCER	IDE	30-10-2	100	**4.08.E-01**	**1.85.E-05**
	DE	30-10-2	100	4.44.E-01	1.29.E-01

5. Conclusion

In this paper, we propose an improved DE algorithm and conduct the performance evaluation through CEC2014 benchmark problems. As a result of the performance evaluation, we confirmed that the overall optimization performance was improved compared with the conventional DE algorithm. We applied the proposed algorithm to the optimization technique of the training multilayer perceptron. In particular, we used a multilayer perceptron with a size of weights approaching 1,000. The performance evaluation shows that the classification performance is improved as compared with the conventional DE algorithm. As a result, we propose an improved DE algorithm that is useful for large scale optimization problems. We extend the proposed algorithm to develop algorithms that can be used to the optimization technique of training Deep learning as a future work.

Acknowledgments

This research was supported by Basic Science Research Program through the National Research Foundation of Korea(NRF) funded by the Ministry of Education(NRF-2015R1D1A1A02062017). Correspondence should be addressed to Dr. Chang Wook Ahn.

References

[1] Rosenblatt, Frank. x. Principles of Neurodynamics: Perceptrons and the Theory of Brain Mechanisms. Spartan Books, Washington DC, 1961

[2] Rumelhart, David E., Geoffrey E. Hinton, and Ronald J. Williams. Learning internal representations by error propagation. No. ICS-8506. CALIFORNIA UNIV SAN DIEGO LA JOLLA INST FOR COGNITIVE SCIENCE, 1985.

[3] Ilonen, Jarmo, Joni-Kristian Kamarainen, and Jouni Lampinen. "Differential evolution training algorithm for feed-forward neural networks." Neural Processing Letters 17.1 (2003): 93-105.

[4] Du, Ji-Xiang, et al. "Shape recognition based on neural networks trained by differential evolution algorithm." Neurocomputing 70.4 (2007): 896-903.

[5] Slowik, Adam, and Michal Bialko. "Training of artificial neural networks using differential evolution algorithm." 2008 Conference on Human System Interactions. IEEE, 2008.

[6] Slowik, Adam. "Application of an adaptive differential evolution algorithm with multiple trial vectors to artificial neural network training." IEEE Transactions on Industrial Electronics 58.8 (2011): 3160-3167.

[7] Storn, Rainer, and Kenneth Price. "Differential evolution–a simple and efficient heuristic for global optimization over continuous spaces." Journal of global optimization 11.4 (1997): 341-359.

[8] Storn, Rainer, and Kenneth Price. Differential evolution-a simple and efficient adaptive scheme for global optimization over continuous spaces. Vol. 3. Berkeley: ICSI, 1995.

[9] Price, Kenneth, Rainer M. Storn, and Jouni A. Lampinen. Differential evolution: a practical approach to global optimization. Springer Science & Business Media, 2006.

[10] Liang, J. J., B. Y. Qu, and P. N. Suganthan. "Problem definitions and evaluation criteria for the CEC 2014 special session and competition on single objective real-parameter numerical optimization." Computational Intelligence Laboratory, Zhengzhou University, Zhengzhou China and Technical Report, Nanyang Technological University, Singapore (2013).

[11] Street, W. Nick, William H. Wolberg, and Olvi L. Mangasarian. "Nuclear feature extraction for breast tumor diagnosis." IS&T/SPIE's Symposium on Electronic Imaging: Science and Technology. International Society for Optics and Photonics, 1993.

IOP Conf. Series: Journal of Physics: Conf. Series **806** (2017) 012011 doi:10.1088/1742-6596/806/1/0120

An Improved SoC Test Scheduling Method Based on Simulated Annealing Algorithm

Jingjing Zheng[1], Zhihang Shen[1], Huaien Gao[1], Bianna Chen[1], Weida Zheng[1] and Xiaoming Xiong[1,2]

[1]GuangDong University of Technology, China
[2]Guangzhou National Integrated Circuited Design Industrialization Center for Modern Service Industry

dayzjj1989123@163.com

Abstract. In this paper, we propose an improved SoC test scheduling method based on simulated annealing algorithm (SA). It is our first to disorganize IP core assignment for each TAM to produce a new solution for SA, allocate TAM width for each TAM using greedy algorithm and calculate corresponding testing time. And accepting the core assignment according to the principle of simulated annealing algorithm and finally attain the optimum solution. Simultaneously, we run the test scheduling experiment with the international reference circuits provided by International Test Conference 2002(ITC'02) and the result shows that our algorithm is superior to the conventional integer linear programming algorithm (ILP), simulated annealing algorithm (SA) and genetic algorithm(GA). When TAM width reaches to 48,56 and 64, the testing time based on our algorithm is lesser than the classic methods and the optimization rates are 30.74%, 3.32%, 16.13% respectively. Moreover, the testing time based on our algorithm is very close to that of improved genetic algorithm (IGA), which is state-of-the-art at present.

1. Introduction

In recent years, higher requirements of size, power consumption, cost, and development cycle on integrated circuit chip have been put forward in various fields such as consumer electronic, and military electronic systems. It is vital that various kinds of functional modules in system are integrated into one chip so the system on chip (SoC) can be built by reusing Intellectual Property (IP) [1]. The number of data ports of IP cores in SoC are different. Therefore, it becomes increasingly difficult to test each IP core when the test access mechanism (TAM) width is limited. The purpose of SoC test scheduling is to shorten the testing time by assigning IP cores to each TAM reasonably [2]. However, test scheduling, which is a Polynomial non-deterministic (NP) problem [3], has reached a bottleneck with the development of the SoC and has aroused widespread concern in the academic and industrial circles. Many academic organizations and international organizations solved this problem so far: the Test Technology Committee of IEEE's computer society set up the P1500 IEEE standard. It became a formal standard in 2005 for a unified IP test interface between the IP core provider and the user [4]; V. Iyengar et al [5] regarded the test scheduling as a ILP (Integer Linear Programming) problem to find an optimal method for SoC test; J. Yang et al [6] combined the representation of closed graph transfer and simulated annealing algorithm to optimize the test scheduling; J. Jia et al [3] described the feasible solution set to test scheduling problem as a population based on the adaptability of genetic algorithm and provided an approximately optimal solution set according to the principle of survival of the fittest evolution; X. Cui et al. [7] proposed a SoC test scheduling method based on ant colony algorithm with the peak temperature constraint. It is obvious that the test time is optimized under the conditions of chip's thermal safety.

In this paper, an improved SoC test scheduling method based on simulated annealing algorithm (SA) by optimizing IP core assignment for each TAM is proposed. Experimental results show that this algorithm is superior to conventional integer linear programming algorithm (ILP) [5], simulated annealing algorithm (SA) [6] and genetic algorithm (GA) [3] in shortening testing time. In addition, the testing time based on our algorithm is very close to that of improved genetic algorithm (IGA) [8], which is state-of-the-art at present.

The remainder of this paper is organized as follows. Section 2 reviews traditional partition problem of SoC testing optimization proposed by V. Iyengar. Section 3 provides a different partition problem from the traditional one and it is solved by an improved simulated annealing algorithm. Experimental results on benchmark circuits are then shown in Section 4. Finally, Section 5 concludes this paper.

2. The traditional partition problem of SoC testing optimization

Problem definition: a testing structure based on wrapper is proposed in IEEE 1500 standard. The wrapper is an interface of combining cores with TAMs, which provides core's normal function and utilizes TAM to test cores and accesses data. In order to effectively solve the combinatorial optimization between the wrapper and TAM, V. Iyengar divided this problem into four sub problems:P_W,P_{AW},P_{NAW}, and P_{NPAW} depending on combinatorial optimization [5], and it can be seen clearly that the previous problem is sub problem of the next one.
Table 1 Four questions proposed by V. Iyengar

Table 1 Four questions proposed by V. Iyengar

P_W	Design a wrapper for a given core, such that the core testing is minimized.
P_{AW}	Determine (i) an assignment of cores to TAMs of given widths and (ii) a wrapper design for each core, such that SOC testing time is minimized. (Item (ii) equals P_W.)
P_{NAW}	Determine (i) a partition of the total TAM width among the given number of TAMs, (ii) an assignment of cores to the TAMs, and (iii) a wrapper design for each core, such that SOC testing time is minimized. (Items (ii) and (iii) together equal P_{AW}.)
P_{NPAW}	Determine (i) the number of TAMs for the SOC, (ii) a partition of the total TAM width among the TAMs, (iii) an assignment of cores to TAMs, and (iv) a wrapper design for each core, such that SOC testing time is minimized. (Items (ii), (iii) and (iv) together equal P_{NAW}.)

Generally speaking, the P_W can be solved well by BFD(best fit decreasing) heuristic algorithm [5]. For problem P_{AW} and P_{NAW}, they can be converted into ILP problem or solved by heuristic algorithms (e.g, simulated annealing algorithm, genetic algorithm, ant colony algorithm). However, These algorithms have shortcomings respectively, i.e. the testing time is too long or the algorithm is too complex. For P_{NPAW}, we generally adopt the method of exhaustion [9].To improve previous effort, this paper study the four problems differing from V. Iyengar's angle.

3. Proposed Algorithm

We notice that , given a fixed core assignment for each TAM, it is easy to determine close-to-optimal TAM widths for each TAM with heuristic algorithms [9], and the problem is transformed into how to search a best core assignment for each TAM. Therefore, the previous four problems proposed by V. Iyengar are rewritten as follows:

① P_W^* :Given a core, design a wrapper such that the core testing time is minimized.

② P_{AW}^*:Given total TAM width W, N_C cores in a SoC, the number of TAMs B and a core assignment for each TAM, determine TAM widths for each TAM such that the SoC testing time is minimized.

③ P_{NAW}^*:Given total TAM width W, N_C cores in a SoC, the number of TAMs B, determine (i) a core assignment for each TAM and (ii) TAM widths for each TAM, such that SoC testing time is minimized.

④ P_{PNAW}^*:Given total TAM width W and N_C cores in a SoC, determine (i) the number of TAMs B, (ii) a core assignment for each TAM and (iii) TAM widths for each TAM, such that SoC testing time is minimized.

Similarly, the four rewritten problems arc progressive, where the former is a sub-problem of the latter. P_W^* is the same as that V.Iyengar defined, which can be solved well with BFD heuristic algorithm [5]. We leave this untouched and denote the solution to P_W^* as Design_wrapper().

3.1 Solution to P_{AW}^* based on greedy algorithm

We define the function solving P_{AW}^* as allocate_TAM_Width(). As this function will be called repeatedly when solving P_{NAW}^*, the running time of allocate_TAM_Width() needs to be very short and we propose a solution based on greedy algorithm, shown as follows. Firstly, we assign one-bit wire to each TAM and calculate the testing time of each TAM using Design_wrapper(). The TAM with the longest testing time is recorded as bottleneck_TAM. And then increasing the width of bottleneck_TAM by one until the total width is run out.

1. Allocate one bit width to every TAM;
2. Repeat when the sum of TAM width less than W;
3. Calculate the testing time of each TAM T(i) (i=1,2,…,B) using Design_wrapper();
4. Record the TAM whose testing time is the longest as bottleneck_TAM;
5. Increase the width of bottleneck_TAM by one;
6. Output TAM width allocation;

To demonstrate the effectiveness of the proposed greedy algorithm, we compare it with the enumerate algorithm. The experiment is based on ITC'02 benchmark SoCs: d695 [10,11]. To simplify the experiment, and without loss of generality, we set the number of TAM to 3. Given total TAM width W and a core assignment, then we can attain the corresponding results shown as table 2, where the solution {(1,6,8),(2,3,4,7),(5,9,10)} means core 1, core 6 and core 8 are assigned to TAM 1 and so on. The table shows that the testing time based on the proposed algorithm and the enumerate algorithm are the same, demonstrating the effectiveness of our proposed algorithm.

Table 2 test result demonstrating the effectiveness of the proposed greedy algorithm

Total TAM width W	Core assignment	The result based on the proposed algorithm	The result based on the enumerate algorithm
16	{(1,6,8),(2,3,4,7),(5,9,10)}	42568	42568
24	{(1,2,10),(3,4,8),(5,6,7,9)}	28289	28289
32	{(1,6,7,10),(2,5),(3,4,8,9)}	21566	21566
40	{(1,5,10),(2,6,7,9),(3,4,8)}	17901	17901
48	{(1,2,6,7,9),(3,4,8),(5,10)}	16984	16984
56	{(1,3,4,8),(2,5,7,10),(6,9)}	13637	13637
64	{(1,2,6,9),(3,4,8),(5,7,10)}	12941	12941

3.2 Solution to P^*_{NAW} and P^*_{PNAW} based on SA

The name and inspiration of simulated annealing algorithm come from annealing in metallurgy. First the solid is heated hot enough. As the temperature of the solid goes up, the thermodynamic free energy of the solid increases and inner particles becomes disordered. Then the solid cools down slowly and meanwhile the inner particles become ordered and stay equilibrium state at every temperature. When the temperature decreases to the minimum, the thermodynamic free energy becomes the minimum [12, 13]. In our proposed SA, the testing time represents the thermodynamic free energy E and the parameter t represents the temperature T. The algorithm flow chart is shown as figure 1.

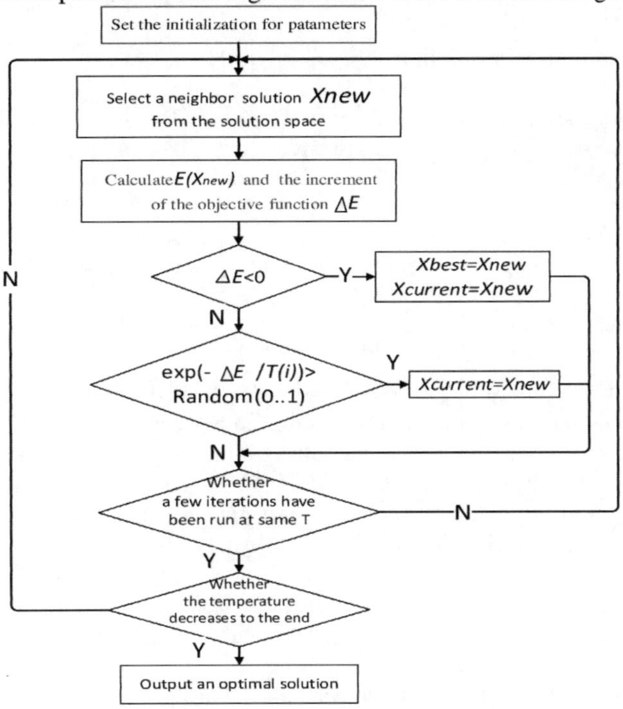

Figure 1. Flow chart of SA.

The main flow of the proposed algorithm is shown as follow. In the beginning, we set the minimum number of TAMs (TAM_Nun_{min}) to 1 and the maximum number of TAMs (TAM_Nun_{max}) to $\min\{N_C, W\}$. Then given number of TAMs, beginning with a random initial core assignment, and keep on searching for its neighbor solutions. Once a feasible solution is obtained, TAM width allocation is performed by allocate_TAM_Width() and the testing time are computed. The new solution will be accepted or abandoned according to the principle of SA. Finally, the best solution is provided during the stochastic search process.

1.$TAM\_Num_{min} = 1$, $TAM\_Num_{max} = \min\{N_C, W\}$;
2.for $TAM_{Num} = TAM\_Num_{min}$ to TAM_Num_{max}
3. create a random initial core assignment X_{new};
4. call allocate_TAM_Width() to perform TAM width allocation and calculate initial testing time E_{new}
5. $E_{best} = E_{new}$, $E_{current} = E_{new}$;
6. set initial temperature $T = T_0$
7. while ($T >$ end temperature T_{end}) ;
// Run a few iterations at same T
8. for each iteration

randomly move to reach a new core assignment X_{new};
call allocate_TAM_Width () to perform TAM width allocation and calculate new testing time_E_{new};
if $E_{new} < E_{current}$ or $e^{\Delta E/T} > rand()$
    $E_{current} = E_{new}$
    update core assignment solution;
  if $E_{new} < E_{best}$
      $E_{best} = E_{new}$
  else
      record the best solution
else
    restore old solution
decrease temperature T at a certain rate α
output the best solution;

3.3 Core assignment based on SA

To avoid repetitiveness of assignment and enhance code efficiency, we have two rules in generating a new random solution:

1. TAM without cores is not allowed;
2. The smallest core index assigned to TAM i is smaller than TAM j, provided i is smaller than j. With the rule, the solution space shrinks to $\frac{1}{B!}$, where B is the number of TAMs;

Following the aforementioned rule, a new solution in SA can be provided by reassignment, which means picking one core from a random set and putting it into another random set. We define this move as M1. To prove the completeness of the above M1, we have a lemma as follow:

Lemma: Given an arbitrary solution A_1,A_2,...,A_B, we are able to reach any other solution A_1^*,A_2^*,...,A_B^* after running M1 finite times.

Proof

1. First we use M1 repeatedly to select the cores of A_1^* from A_2,...,A_B and put them into A_1 until all the cores of A_1^* are in A_1.(Once a set is empty, we can select a core which does not belong to A_1^* from other sets and put it into the set)
2. Similarly, we are able to reach that all the cores of A_i^* are in A_i. ($i = 1,2 \ldots B$)

We assume that the number of cores in A_i^* is Core_Num_i^*, the number of cores in A_i after finite M1 is Core_Num_i. And the total number of cores is N_C, that is $\sum_1^B \text{Core\_Num}_i^* = N_C$. Based on step 2, we have $A_i^* \subset A_i$, that is $\text{Core\_Num}_i \geq \text{Core\_Num}_i^*$ ($i = 1,2 \ldots B$). After limiting M1, the total number of cores does not change. Consequently, $\sum_1^B \text{Core\_Num}_i = \sum_1^B \text{Core\_Num}_i^* = N_C$. The above equation is true only when $\text{Core\_Num}_i = \text{Core\_Num}_i^*$. Therefore, we are able to reach that $A_i = A_i^*(i = 1,2 \ldots, B)$.

4. Experimental result

The Comparison among the proposed algorithm and the traditional ILP algorithm [5], simulated annealing algorithm, and genetic algorithm was taken using the ITC'02 benchmark SoC: d695.

Results are shown in table 3 and figure 2. Table 3 shows the testing time of each algorithm when the total TAM width is 16, 24, 32, 40, 48, 56, 64 respectively. By contrast, the testing time by our algorithm is basically less than other classic and state of the art methods. Moreover, Figure 3 shows that the advantage is more obvious when TAM width reaches larger. When TAM width reaches 48,56 and 64, the testing time by our algorithm is lesser than that by others and the optimization rate are 30.74%, 3.32% and 16.13% respectively.

Table 3. The test result of conventional algorithms and ours

benchmark SoC	algorithm	Total TAM width						
		16	24	32	40	48	56	64
		Testing /Reference clock cycle						
D695	ILP	42952	28327	21423	17210	16975	16516	15694
	SA	42361	28638	21967	17414	15141	12692	11242
	GA	41949	28290	21329	17084	16975	14610	12960
	Our algorithm	42268	28289	21437	17523	14310	12462	10869

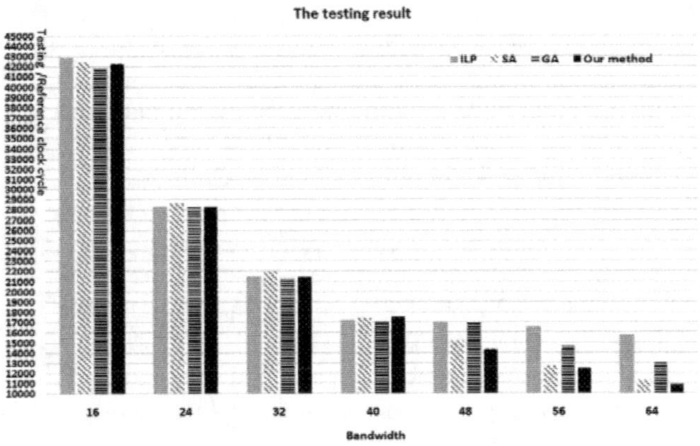

Figure 2. The comparison result with traditional algorithms

To further verify the validity of the proposed method, a comparison with the state-of-the-art improved genetic algorithm was taken [8]. With the same experimental setting, the comparing results is shown in Figure 3. The results shows that the testing time by our method is very close to that by IGA algorithm. This difference becomes smaller when the total TAM width increases. When the TAM width reaches to 48, 56 and 64, the testing time by our algorithm is 14310/12462/10869 and that by IGA is 14310/12540/10571, that is, the rate of optimization is 0/0.62% / -2.82%, respectively.

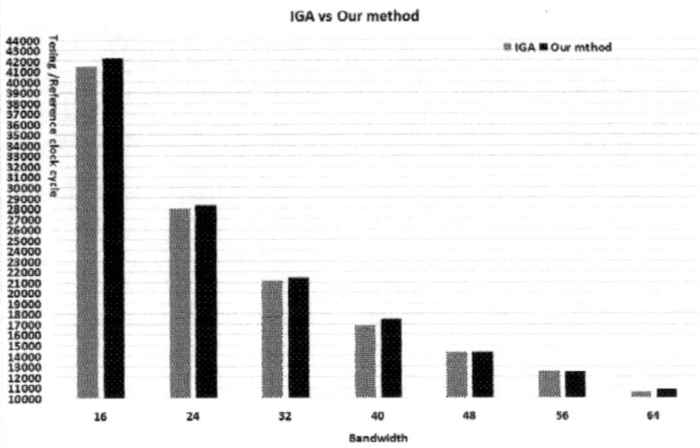

Figure 3. The comparison result with IGA

5. Conclusion

This paper proposed an improved simulated annealing algorithm to solve the SoC test scheduling by searching an optimal core assignment to each TAM. In term of core assignment based on SA, to prove the completeness of the M1, we defined a lemma and given it proof. The experimental result shows much better performance comparing with the classic ILP/SA/GA algorithm and is very close to that of state-of-the-art IGA. Our further work can be done by implementing this algorithm in considering constrains of the power consumption.

Reference

[1] C. P. Yue, D. Luo, G. Zhu, Y. Wang, K. Q. Maqbool and Z. Li, Recent developments in transceiver SoC design for next generation optical networks[C], *2015 Asia-Pacific Microwave Conference (APMC)*, Nanjing, 2015, pp. 1-3.

[2] Lu Xi and Mo Tingting. Parallel Test Task Search Genetic Algorithm[J]. *Microelectronics & Computer*, 2015,03:146-150.

[3] Lei Jia and Fang Gang. An Test Scheduling Method of SoC based on Genetic algorithm[J]. *Chinese Journal of Scientific Instrument*, 2007 28（4）:15—18.

[4] IEEE Std 1500-2005.IEEE 1500 Standard For Embedded Core Test[S].

[5] V. Iyengar, K. Chakrabarty and E. J. Marinissen, Test wrapper and test access mechanism co-optimization for system-on-chip[C], Test Conference, 2001. Proceedings. International, Baltimore, MD, 2001, pp. 1023-1032.

[6] Yang Jun and Luo Lan. SoC test schedule based on TCG and simulated annealing algorithm[J]. *Journal of Circuits and Systems*, 2006,05:37--43.

[7] Cui Xiaole, Xiong Zhitian, Cheng Wei and Li Chongren. Thermal-aware SoC test scheduling method based on ant colony optimization[J]. *Chinese Journal of Scientific Instrument*, 2014, 04:948-953.

[8] Liu Wei and Li Bin. An Improved Test Scheduling Method of SoC Based on Genetic Algorithm. *Radio Communications Technology*, 2016,42(2) : 37－40.

[9] Li Jiang, L. Huang and Q. Xu, Test architecture design and optimization for three-dimensional SoCs[C], *2009 Design, Automation & Test in Europe Conference & Exhibition*, Nice, 2009, pp. 220-225.

[10] E. J. Marinissen, V. Iyengar and K. Chakrabarty, A set of benchmarks for modular testing of SOCs[C], *Test Conference, 2002. Proceedings. International*, 2002, pp.519-528.

[11] H. Zhang and V. D. Agrawal, SoC TAM Design to Minimize Test Application Time[C], *Test Workshop (NATW), 2015 IEEE 24th North Atlantic*, Johnson City, NY, 2015, pp. 55-60.

[12] Xie Yun. Principle and Realization of the Simulated Annealing Algorithm, *Numerical Mathematics A Journal of Chinese Universities*. 1999,03:212-218.

[13] W. Wu, L. Li and X. Yao, Improved simulated annealing algorithm for task allocation in real-time distributed systems[C], *Signal Processing, Communications and Computing (ICSPCC), 2014 IEEE International Conference on*, Guilin, 2014, pp. 50-54.

The Design and Implementation of a Low Cost and High Security Smart Home System Based on Wi-Fi and SSL Technologies

Chong-Yao Xu[1], Xin Zheng[1] and Xiao-Ming Xiong[1,2]

[1] Guangdong University of Technology, School of Automation, No. 100 Waihuan Xi Road Guangzhou Higher Education Mega Center Panyu District, 510006 Guangzhou P.R China

[2] Guangzhou National Integrated Circuit Design Industrialization Center for Modern Service Industry

chongyaoxu@126.com, 18963983546@163.com, xmxiong@gdut.edu.cn

Abstract. With the development of Internet of Things (IoT) and the popularity of intelligent mobile terminals, smart home system has come into people's vision. However, due to the high cost, complex installation and inconvenience, as well as network security issues, smart home system has not been popularized. In this paper, combined with Wi-Fi technology, Android system, cloud server and SSL security protocol, a new set of smart home system is designed, with low cost, easy operation, high security and stability. The system consists of Wi-Fi smart node (WSN), Android client and cloud server. In order to reduce system cost and complexity of the installation, each Wi-Fi transceiver, appliance control logic and data conversion in the WSN is setup by a single chip. In addition, all the data of the WSN can be uploaded to the server through the home router, without having to transit through the gateway. All the appliance status information and environmental information are preserved in the cloud server. Furthermore, to ensure the security of information, the Secure Sockets Layer (SSL) protocol is used in the WSN communication with the server. What's more, to improve the comfort and simplify the operation, Android client is designed with room pattern to control home appliances more realistic, and more convenient.

1. Introduction

Recently, the Internet of Things (IoT) and intelligent hardware have shown explosive growth, and the intelligent terminals can be seen everywhere. On the one hand, home appliances can access to the Internet through the Wi-Fi, thus it can be accurately controlled and monitored by a mobile phone or a tablet, which greatly improve people's living quality and ensure the safety of home [1-4]. On the other hand, as the IoT technology become more sophisticated, with the continuously decreasing cost and more convenient operation of smart home, the popularity of smart home system has become an inevitable trend.

1.1. Existing smart home technologies

As early as 1984, the United States had the world's first smart home building. After decades of development, the smart home system has experienced quite a few great reforms. Reference [5] introduces a smart home system based on Java, all of the intelligent appliances are connected together with physical means, and through a personal computer (PC) to realize the separate control, with providing a secure network interface. However, due to the complex installation of physical connections, and adding the personal PC and physical connections, the cost of system implementation is very expensive. Reference [6] proposed a smart home system based on Bluetooth (BT), each home appliances are equipped with a local BT sub-controller, which unified the data sending to the BT master controller. The scheme solves the problem of complex cable installation and connection, but it is constrained by the short distance of BT transmission, the limitation of the network and the lack of remote control. Reference [7-8] introduced a smart home system that controls appliances through global system for global system for mobile communications (GSM) and general packet radio service (GPRS) .The program is not subject to distance restrictions, and the user can control it through the mobile phone. However, because the user has to remember the control code of the appliance and the telephone number corresponding to each appliance, each appliance requires a GSM/GPRS module with high cost, and each module needs to install the Subscriber Identification Module (SIM) card separately. Reference [9] presented a kind of smart home system based on ZigBee network, which solves the limitation of the BT network. Because the ZigBee can't directly access to the Internet or communicate with mobile phone, so in order to realize the remote control, it is necessary to design an embedded development board with Ethernet interface or Wi-Fi functions as a gateway, as a centralized controller achieving unified transfer of data. Reference [10] described a solution for controlling television through gestures, which means the television can be controlled by recognizing different gestures of the user. It consists of a set of ultrasonic sensor, a red/green/blue (RGB) camera, and a depth camera above the television. Combined with the recognition algorithm, it has unique advantages. The deficiency of the method is that it is susceptible to gestures accuracy and lighting. Besides, it's easy to cause fatigue when the user repeatedly making gestures [9]. Reference [11] introduced a smart home control system based on the ZigBee wireless sensor networks and power line communications. Each electrical equipment in the system is equipped with a power line communication transceiver, which receives control information from management station by the power line to control appliances and to feedback the appliance status to the management system. Each room has multiple ZigBee sensor nodes and a power line communications (PLC) transceiver integrating a ZigBee coordinator which sends data to the management system after receiving the information of sensor node through the PLC transceiver. Compared with simple Zigbee system, the system can effectively improve the data accuracy of sensor nodes in a large amount of data (>200Kbytes/s), which has good innovation. In addition, many foreign researchers has proposed many power management systems of smart home [12-15], and reference [15-17] shows that intelligent residence can be more efficient than traditional residential power utilization.

1.2. The characteristics and innovation of the system

The improvements and innovations of the proposed system are described as following:

1) All Wi-Fi Smart Nodes (WSNs) are accessed to the Internet through the router, and are directly controlled by mobile phone, without centralizing control of data transfer. It can reduce the complexity of installation and debugging, enhance the scalability, and decreases the cost drastically.
2) ESP8266, with high performance and low cost Wi-Fi solution, is used in the smart appliances node. It is a system on a chip (SOC) design, including an internal 32-bit embedded professor, a Wi-Fi transceiver, an analog-to-digital converter (ADC), input/output (I/O) controllers, serial transceivers and so on. There is no need to add other microcontroller unit (MCU), so it can improve the stability and reduce the cost of the system.
3) WSN module can be embedded with a Secure Sockets Layer (SSL) to encrypt the data, which can ensure the information security of remote control, and reduce the risk of system attacks.
4) The Application (APP) based on Android client is designed according to a pattern of room, and all appliances have a real simulation effect that improving the operability of the system.

Moreover, users can also add the intelligent Wi-Fi nodes to increase home appliances through APP which can increase the flexibility and extendibility of the smart home system.

2. System architerture

The smart home system mainly includes three parts: WSN, Cloud Sever and APP Client. The system architecture is shown in figure 1. Each household application is equipped with a WSN which can access to the Internet through Wi-Fi network. A database is built on the cloud server to record status of home appliances, environment conditions and user's habits. Users can control every home appliance or detect the environment when you login in the server at anywhere. The SSL security protocol is embedded in the WSN, which is used to encrypt data between cloud server and WSN with 1024 bit keys.

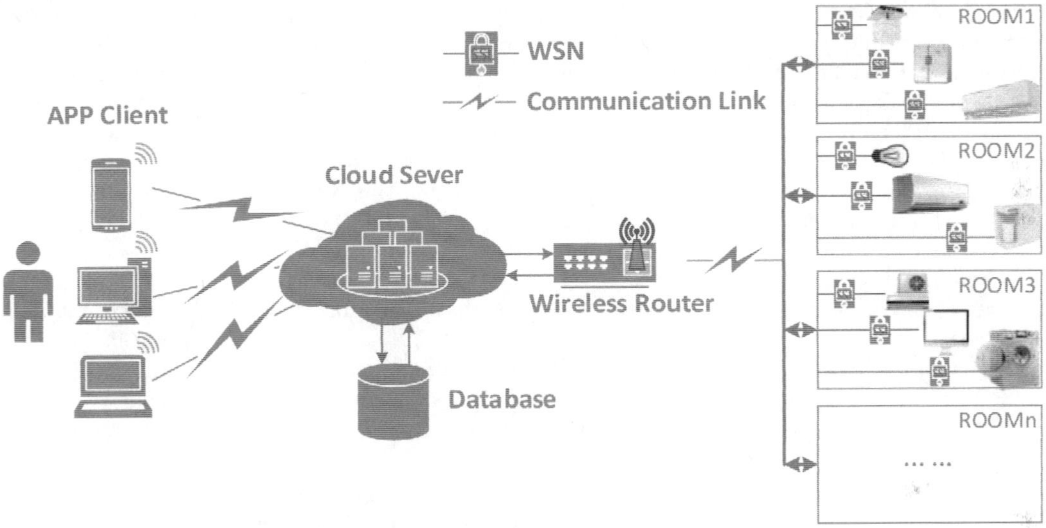

Figure 1. System architecture overview

2.1. WSN

Wi-Fi technology is a popular wireless communication protocol, because of its high communication speed and easy operation, it has a wide range of applications. Now almost every smartphone has Wi-Fi configuration, and Wi-Fi hotspots (APs) are covered in the home, school, and every prosperous business district. In recent years, with the rapid development of the IoT technology, the cost of the Wi-Fi chip decreases gradually and the function is becoming stronger. It has the advantages such as strong versatility, high speed, simple operation and low cost, using Wi-Fi as the wireless solutions of smart home system. It should be noted that the retail price of Wi-Fi chip for this design is less than 0.88 USA dollar (USD), which is programmable and can be used as MCU.

In this design, each WSN is integrated with Wi-Fi functions, which can be accessed directly through the router to the cloud server. The socket, fluorescent lamp, TV and other appliances are connected to the same Wi-Fi network via WSN, so as to set up a home Wi-Fi network, users can access through the cloud server to control appliances and realize the environment situation in the home.

2.2. Cloud Server

Cloud server is also known as cloud computing server, which has three core elements: computing, network and storage. It is based on the cluster server technology. Cloud server is simple, efficient, safe and reliable. It has flexible and scalable processing capacity. It has four main features including high-density, energy-saving, easy reorganization, and system optimization. Appling cloud sever into the smart home system can bring more safety and stability to the smart home system.

The cloud server we used is a free Internet cloud platform offered by a third party companies. It mainly helps users set up the remote control of home appliances. When users need to control home appliances or learn the home situation, they only open the APP to login into the server and click on the control button of appliance, then the APP will automatically upload the information to the server which sends control commands to the corresponding home appliance and feedback control results to the client after the server receives the data.

2.3. App Client

As technology evolves, mobile phones, Pad and other portable intelligent terminals become more pervasive, almost everyone will take at least one intelligent terminal. Intelligent mobile terminal is selected as the control terminal of smart home system, which can not only reduce the complexity of the system, improve the flexibility of control and ease of operation, but also greatly reduce the cost of the system. Android system, as one of the three major mobile operating systems, it is favored by customers and mobile phone manufacturers for its completely free, open source code, plentiful application software features. Therefore, in this paper, a control software based on Android platform is designed for realizing the function of all kinds of control. The software has a warm interface, a reasonable layout, powerful functions and can be expanded flexibly.

3. System implementation

The photograph of the proposed smart home system is shown in figure 2. The realization of the design of the WSN mainly consists of Wi-Fi smart socket, Infrared Radiation (IR) learning module, CO2 sensor, harmful gas (CO) monitoring module. Router is used as AP to launch wireless signal to build home Wi-Fi network. When WSNs are connected to the router, using the mobile phone to login to server, the system will work properly.

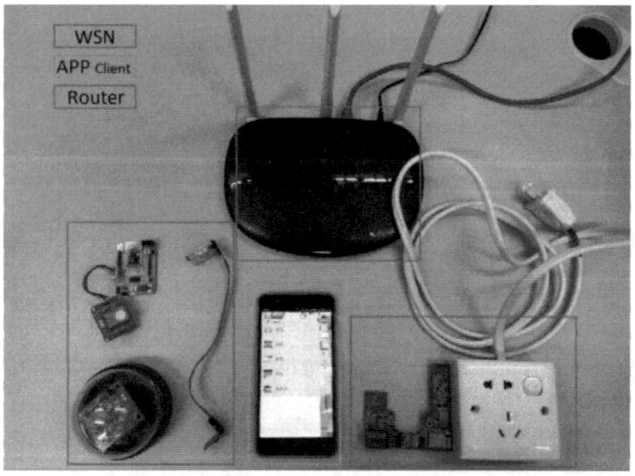

Figure 2. System implementation

3.1. WSN hardware design

In this system, we designed four WSNs to comprise a simple smart home system, including smart socket, IR learning module, CO2 sensor module and harmful gas (CO) monitoring module. The detailed information will be introduced as following.

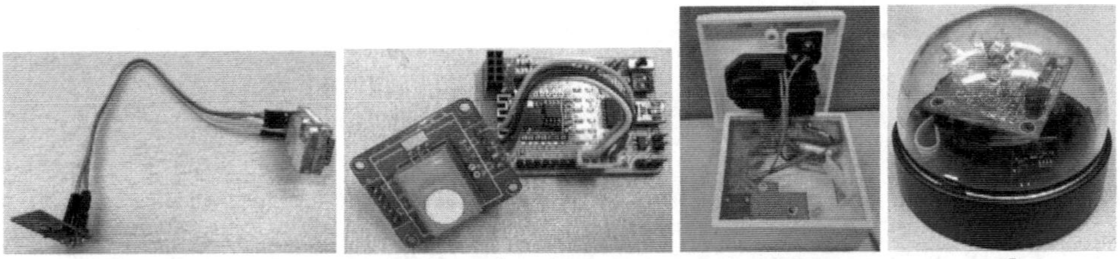

a. b. c. d.

Figure 3. (a) CO2 sensor module **(b)** harmful gas (CO) monitoring module **(c)** smart socket **(d)** IR learning module

1) CO2 sensor module: This module is composed of CO2 concentration sensor and Wi-Fi module, as shown in figure 3 (a). Wi-Fi module read the data a sensor every ten minutes via the serial port, and feedback to the server, then enter into the low power mode with only 12uA current consumption. Ten minutes later, the Wi-Fi module will automatically wake up and read the next data.
2) Harmful gas (CO) monitoring module: figure 3 (b) illustrates the printed circuit boards (PCBs) used in the system debugging, this module is mainly responsible for monitoring the indoor CO concentration. Wi-Fi module adopts the Modbus protocol to communicate with the sensor through the serial port, query data each one minute, and upload the result to the server, presently enter the low power state. When the concentration of CO exceeds the standard, the module will send an emergency notification to the mobile phone through the server to remind the user.
3) Smart socket: Smart socket products is shown as in figure 3 (c). It is mainly ideal for receiving and transmitting control information through the Wi-Fi chip, In addition, the S8050 low cost transistor is used to driver relays, so it can achieve on and off of the relay. Then the 220V socket can be controlled to handle the appliances work or not.
4) IR learning module: This module is shown as in figure 3 (d), which is mainly used for controlling air conditioner, television and other IR equipment. It has a universal IR receiver and a MCU with Cortex-M3 kernel. The MCU can learn the encoding of all kinds of IR controllers through calculating the elapsed time of high and low voltage of IR encoding. In addition, there are IR emitters in 8 directions to transmit the corresponding signal to control appliances according to the control information sent from the mobile phone.

3.2. Embedded software design

The embedded software designed by this system refers to the internal control program of Wi-Fi chip. The program is mainly responsible for the connection of Wi-Fi network, work mode settings, Wi-Fi data transmission and reception, IO port control, serial data dispatch, AD data acquisition and so on.

WSN will start automatically in Station+AP mode, and try to connect Wi-Fi according to the last recorded information, WSN switches from Station+AP mode to Station mode after successful Wi-Fi connection, if it is not connected in a minute, we can access the Wi-Fi wireless network from WSN itself by mobile phone, and configure Wi-Fi network information (SSID and password) which WSN needs through the APP. Besides, it will automatically connect to the cloud server and enter the suspended state after successful Wi-Fi connection. If the WSN is the sensor node, it will automatically report to the server. The flow chart of the program is shown in figure 4.

3.3. APP design

In this system, an application program based on Android (APP) is designed for home appliances control. In order to facilitate the operation, Android client is designed in room pattern to control home appliances, such as drawing and graphical interface, each appliance has its corresponding graphical interface, as shown in figure 5. Simultaneously, the APP also supports to increase home appliances against actual conditions.

3.4. Control protocol

Control protocol is shown in figure 6. According to the user's requirement, there are 12 bytes control information in the system, the first 2 bytes is fixed for two hexadecimal number--0xA5, 0x5A, followed by 4 bytes of user ID, 1 byte random number, 1 byte room code, 1 byte command code and 2 bytes CRC check code. After the server receives the control information, firstly it returns instructions to confirm the information, and then controls the home appliances, finally it feedback the current status of home appliances and environmental information to the client.

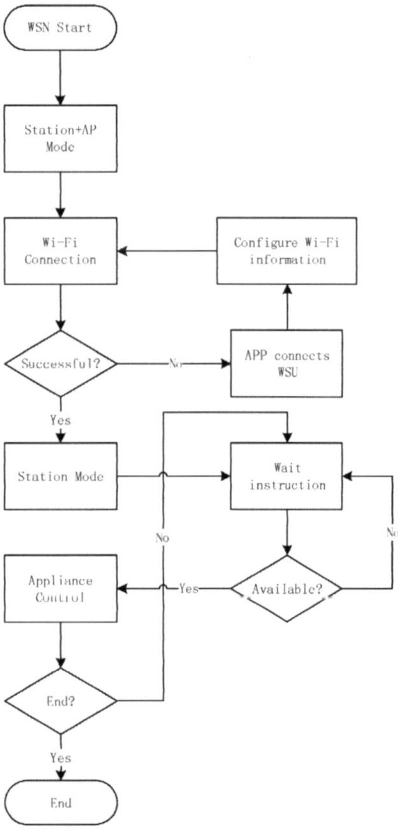

Figure 4. Program flow chart

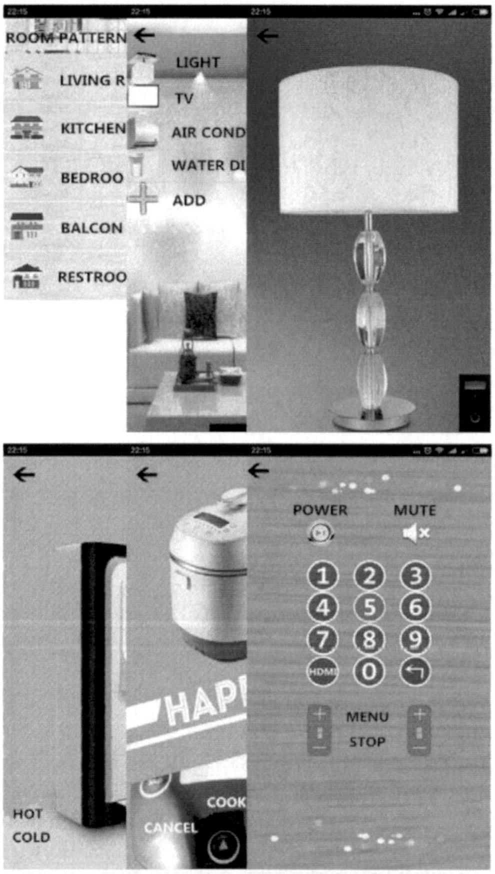

Figure 5. APP control interface

XX X XXXX X X X XX

① ② ③ ④ ⑤ ⑥ ⑦

Figure 6. Control protocol

①→Start flag, two bytes, fixed to 0xA5, 0x5A;

②→Instruction random code, one byte, to ensure that this section of the instructions correspond with the server response information;

③→User ID, four bytes, user unique certification;

④→Room information, one byte, indicates which room the home appliance is controlled by this instruction;

⑤→Appliance information, one byte, the only representation of a home appliance;

⑥→Control instruction, one byte, controls the operation of home appliances. For example: light switch, TV channel selection, air conditioning temperature switch, etc.;

⑦→CRC check, two bytes, verifies whether the transmitted information error, etc..

3.5. SSL security encryption

SSL encryption technology is one of the most securely and widely used network encryption technology. It is between the TCP/IP and many application protocols, providing security support for data communication. The SSL encryption certificate is implanted in the WSN node, so as to construct a secure channel between the WSN and the server to carry out data transmission. Owing to a vast amount of memory required for the SSL function (under SSL cache settings for the 8KB, SSL function requires at least 22KB space [18], and available to users of RAM is about 50KB [19], SSL). To set whether to open the SSL function via the macro definition before compilation, thus ensuring the flexibility of application.

The implementation method of the WSN as the SSL Client is described as following. SSL security encryption using the Software Development Kit (SDK) provided from ESP8266 chip manufacturers. Firstly, apply to Espressif Systems for a certificate synthesis script. Secondly, put it in the same directory as the CA certificate. Thirdly, run certificate synthesis script in lubuntu system so that the certificate will be made into a BIN file. Finally, download the BIN file into the specified location [18]. Certification process is as follows:

I. WSN initiates TCP connection, then read the CA certificate in the Flash, and load it into the corresponding SSL_CTX structure.

II. ESP8266 launches TLS handshake, get the certificate issued by SSL server, and verify CA certificate supported by WSN, if successful, completing the verification, otherwise disconnected.

4. Cost analysis

Because the system is directly connected to the router by WSN and accessed to the Internet without a centralized controller, so it can greatly reduce the cost of the system. Specifically, WSN adopts a single chip solution, the work of Wi-Fi communications and home appliances control are finished by a low cost chip. In this section, he Wi-Fi smart socket is considered as an example to analyze the market advantage of the system. The smart socket mainly composed of Wi-Fi module, relay, driving transistor, 3.3V voltage regulator chip, and the cost list is shown in the following table 1:

Table 1. Price Sheet

Component	**Price (USD $)**
Wi-Fi chip	0.88
32M SPI Flash	0.07
S8050 triode	0.01
Relay	0.18
ASM1117-3.3V	0.03
PCB board	0.29
Power module	0.59
Total	2.05

5. Test result

5.1. Wi-Fi socket

Wi-Fi socket is one of the most frequently used devices in smart home system. It can realize the control of household appliances such as fluorescent lamp, electric fan, water heater, curtain, etc. By repeating testing, the Wi-Fi socket can flexibly turn on and off the light through the APP. figure 7 shows this working process. Considering the time-delay of relay, the light will be turned on after operating APP about 500ms later.

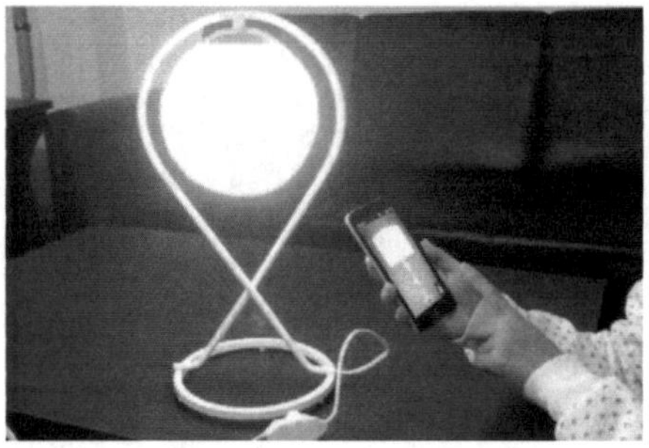

Figure 7. Light control

Figure 8. TV control

5.2. TV control

The intelligent IR learning module is also tested in this paper. Firstly, the self-learning method is used to learn the IR coding for remote controlling TV set in the IR module, and then the different keys are numbered. When a key is pressed, the mobile phone will send the corresponding number through the Wi-Fi network to the IR learning module which sends the corresponding IR signal after received the information, so it can realize the control of the TV.

Figure. 8 shows a video screenshot of controlling TV by APP and IR learning module. The results of practical tests proves that it is easy to achieve all the functions of the remote control, including "program +", "program –", "silent", "channel switch" and so on. In figure. 8, "a" is the mobile phone control terminal; "b" is the remote control; "c" is the IR learning module.

6. Conclusion

A smart home system solution with the properties of low cost, high security and reliability is introduced in this paper. Four kinds of WSN are developed. All WSN nodes can be connected directly to the Wi-Fi system, and access to the server via the SSL security protocol. At the same time, the APP based on the Android is designed for WSN controlling. Each WSN can work independently, and the user can control the nodes directly through the mobile phone without a centralized controller, which avoid the complication of system installation and commissioning process. And due to the low cost of the system, it has great market value and competitive advantage.

References

[1] Kim T, Lee H, and Chung Y 2010 Advanced universal remote controller for home automation and security. *IEEE Trans. Consum. Electron.,* **56(4)** 2537-42.

[2] Pavithra D and Balakrishnan R 2015 IoT based monitoring and control system for home automation. *Communication Technologies (GCCT), 2015 Global Conf. on. IEEE* (India: Kanya Kumari District) pp 169-173.

[3] Wenbo Y, Quanyu W, et al 2015 Smart home implementation based on internet and Wi-Fi technology. *Control Conf. (CCC), 2015 34th Chinese. IEEE* (Hangzhou: Zhejiang University) pp 9072-77.

[4] Yun J, Ahn I Y, Sung N M and Kim J 2015 A device software platform for consumer

electronics based on the Internet of Things. *IEEE Trans. Consum. Electron.*, **61(4)** 564-571.

[5] Al-Ali A R, Al-Rousan M 2004 Java-based home automation system. *IEEE Trans. Consum. Electron.*, **50(2)** 498-504.

[6] Sriskanthan N, Tan F, Karande A. 2002 Bluetooth based home automation system. *IEEE Trans. Consum. Electron.*, **26(6)** 281-289.

[7] ElKamchouchi H, ElShafee A. 2012 Design and prototype implementation of SMS based home automation system. *Electronics Design, Systems and Applications (ICEDSA), 2012 IEEE Int. Conf. on. IEEE* (Malaysia: Kuala Lumpur) pp 162-167.

[8] Wei L, Wei L, et al 2013 Design and Implement on Smart Home System. *Intelligent Systems Design and Engineering Applications (ISDEA), 2013 4th Int. Conf. on. IEEE* (China: Zhangjiajie) pp 229-231.

[9] Gill K, Yang S H, Yao F, and Lu X 2009 A zigbee-based home automation system. *IEEE Trans. Consum. Electron.*, **55(2)** 422-430.

[10] Lian S, Hu W and Wang K 2014 Automatic user state recognition for hand gesture based low-cost television control system. *IEEE Trans. Consum. Electron.*, **60(1)** 107-115.

[11] Li M and Lin H J 2015 Design and implementation of smart home control systems based on wireless sensor networks and power line communications. *IEEE Trans. Ind. Electron.* **62(7)** 4430-42.

[12] Byun J, Hong I, Lee B and Park S 2013 Intelligent household LED lighting system considering energy efficiency and user satisfaction. *IEEE Trans. Consum. Electron.*, **59(1)** 70-76.

[13] Han D M and Lim J H 2010 Design and implementation of smart home energy management systems based on zigbee. *IEEE Trans. Consum. Electron.*, **56(3)** 1417-25.

[14] Han J, Choi C S, Park W K and Lee I 2014 Smart home energy management system including renewable energy based on ZigBee and PLC. *IEEE Trans. Consum. Electron.*, **60(2)** 198-202.

[15] Han D M and Lim J H 2010 Smart home energy management system using IEEE 802.15. 4 and zigbee. *IEEE Trans. Consum. Electron.*, **56(3)** 1403-10.

[16] Missaoui R, Joumaa H, Ploix S and Bacha S 2014 Managing energy smart homes according to energy prices: analysis of a building energy management system. *Energy and Buildings,* **71** 155-167.

[17] Molitor C, Benigni A, et al 2013 Multiphysics test bed for renewable energy systems in smart homes. *IEEE Trans. Ind. Electron.*, **60(3)** 1235-48.

[18] Espressif Systems IOT Team 2015 *ESP8266 SDK SSL User Manual CN v1.1* http://bbs.espressif.com/.

[19] Espressif Systems IOT Team 2015 *ESP8266 Datasheet CN v4.4* http://bbs.espressif.com/.

Pressure Characteristic Analysis of a Hydraulic System

H. Y. Cho[a] and H. J. Yang[b]

[a] Hydraulic Friends Co., Ltd. 1MA-708, Sihwa Industrial Complex Jongwang-Dong, Siheung-Si, Gyeonggi-Do, South Korea
[b] Department of Mechanical Design Engineering, Korea Polytechnic University, Siheung-city, Gyeonggi-Do, 15073, South Korea

Abstract. EPPR(ElectroProportional Pressure Reducing) valve control the MCV(Main Control Valve) built on the mobile heavy machine. The EPPR valve was tested in the experimental setup and the performance of the valve was compared with that of the existing EPPR valve. On thisstudy, electromagnetic properties analysis using AMESim program was performed to optimize the designing of EPPR Valve (Electric Proportional Pressure Reducing Valve) and by applying its results to the hydraulic system analytical model, performance of the valve could be predicted. Also by comparing the results of the actual experiment and the simulation, The results of thisstudy is that the 3 factor(cone angle, tip width, clearance between sleeve and plunger) have much effectiveness than other components in the EPPR valve.

1. Introduction

EPPR valve (Electric Proportional Pressure Reducing Valve) has been used in controlling hydraulic system electrically. It is the valve controlling pressure proportional input-current and is using industrial department (fork lift, excavator, special motorcar, heavy machine) variously [1,2].

As shown in figure 1, the study of EPPR Valve has been very important because it must satisfied fl ow force, hysteresis, linear form, repeating precision. So, it was performed hydraulic characteristic of newdesigned model EPPR valve using of AMESim and compared with experimental method.

2. Modeling for AMESim analysis.

Modeling of EPPR valve was completed with assemble HCD component and library in AMESim program. The viscous friction value of spool was determined two kinds method [3,4].

First, the case of non- spring model, it was used viscous friction coefficient of the meaning of physical.

$$C_v = \frac{\mu(\pi D)L}{clearance} \quad \left[N/(m/s) \right]$$

$$\mu = \rho \upsilon$$

D : diameter of mass
L : contact length of mass and chamber
clearance : distance within of mass and chamber

Second, the case of spring existing model, it was used viscous friction coefficient of the meaning of math ematical.

$$C_v = 2\zeta\sqrt{km} \quad \left[N/(m/s) \right]$$

$$\zeta = damping\ ratio(0.4)$$

On the above expression, the system was damped over and oscillation was not created if damping ratio was too big, the system was damped under and oscillation was created if damping ratio was too small. But the parameter value showing above expression was not just exacted and the various parameter operated to determine viscous friction coefficient. The exact value of viscous friction coefficient was determined only by experiment. But, on this paper, additional experiment was omitted and added next assumption.

· The temperature variation of system was assumed normal condition maintaining initial condition.
· The volume of load side was constant.

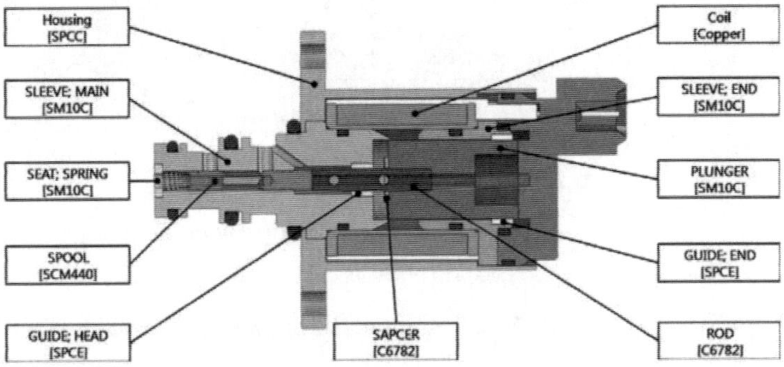

Figure 1. structure of EPPR valve

3. Results of AMESim analysis

The valve of supply pressure Ps was 25bar and viscosity damping coefficient B was 8000 [N/(m/sec)] and another parameter was displayed in table 1. The parameter of solenoid was shown in figure 2 [5], [6]. The parameter value of solenoid valve was displayed in table 2. AMESim model of EPPR Valve was s hown in figure 3. Using of 3D table editor, as shown in figure 4, the magnetic analysis values of Max well program was inputted in AMESim program parameter and the output values of P-I diagram was generated according to spool stroke.

Table 1. parameter value of EPPR Valve

	parameter name	parameter value	unit
M_s	spool mass	0.97	g
M_s	amateur mass	27.75	g
B	viscosity damping coefficient	8000	N/(m/sec)
K_s	spool spring coefficient	0.2353	N/mm
d_s	spool diameter	3.5	mm
β	volume elastic coefficient	8000	bar
P_s	supply pressure	25	bar
P_t	tank pressure	0	bar
C_d	orifice coefficient	0.61	-

ρ	oil density	880	kg/m^3

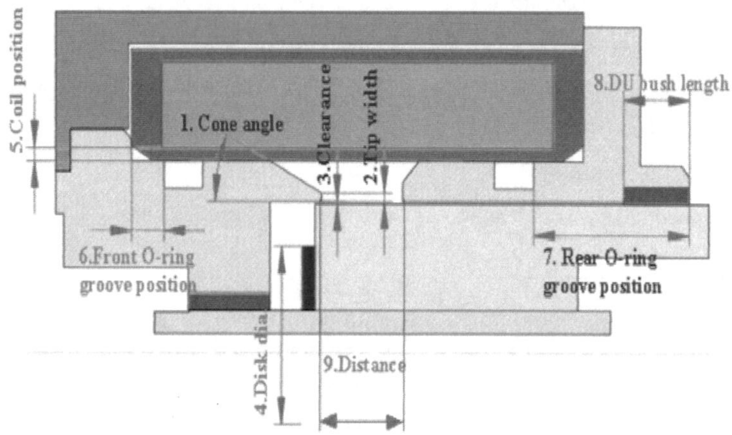

Figure 2. paratemeter of solenoid valve

Table 2. Changing value of solenoid valve shape

	cone angle	tip width	clearance	rear groove	front groove
existent value	27^0	0.3mm	0.2mm	13.8mm	3.6mm
new value	26^0	0.2mm	0.1mm	9.5mm	2.0mm

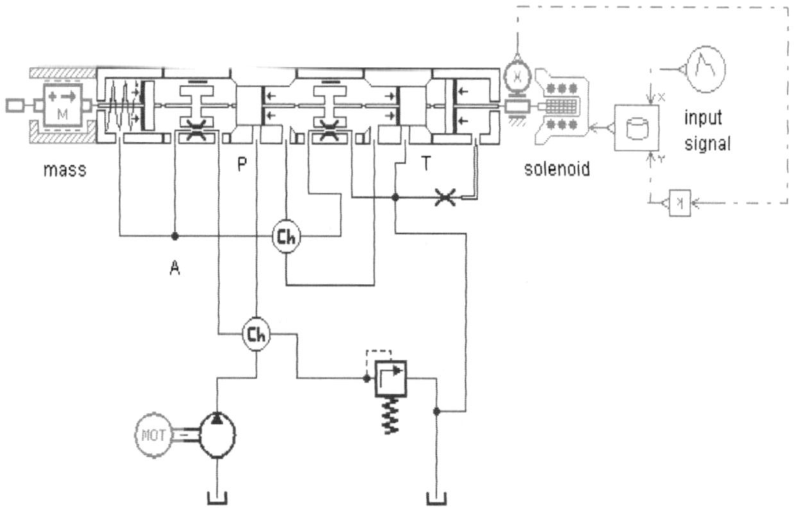

Figure 3. AMESim model of EPPR Valve

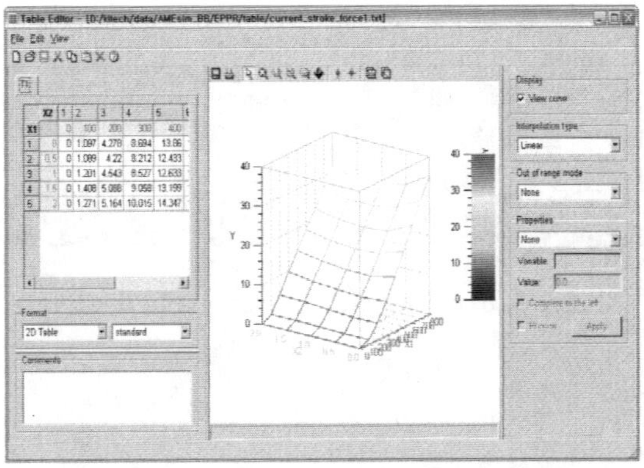

Figure 4. 3D table of current-stroke force

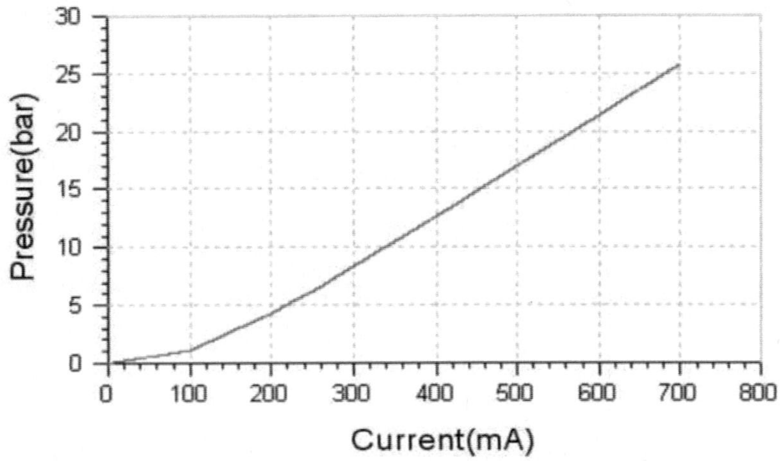

Figure 5. P-I diagram of AMESim analysis result adapting new value

Resulting of AMESim program analysis, the pressure linear form of EPPR valve was in proportion from 100 mA to 700 mA and the range of pressure was controlled from 0 bar to 25 bar as shown in fig ure 5 [7].

4. Hydraulic characteristic experiment of EPPR valve

It was obtained the P-I diagram according to current value of EPPR valve by experiment connecting with PC type data acquisition system and then the supply pressure Ps was 50 bar. The experimental system of EPPR valve was shown in figure 6.

[hydraulic power unit & test stand]

- Regular pressure : 210 bar
- Regular flow : 70 lpm
- 16 channel strain amplifier

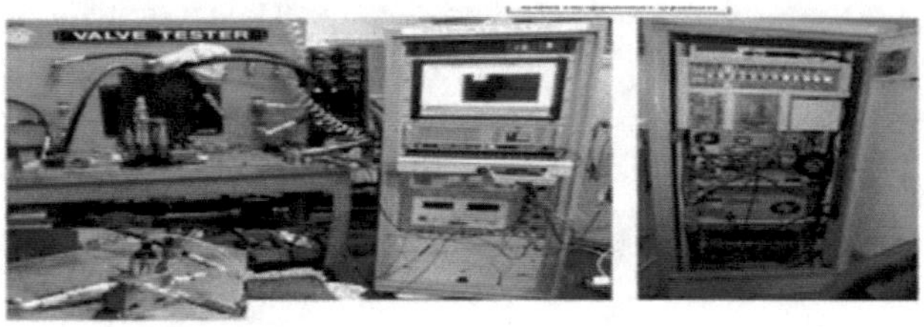

Figure 6. experimental system of EPPR valve

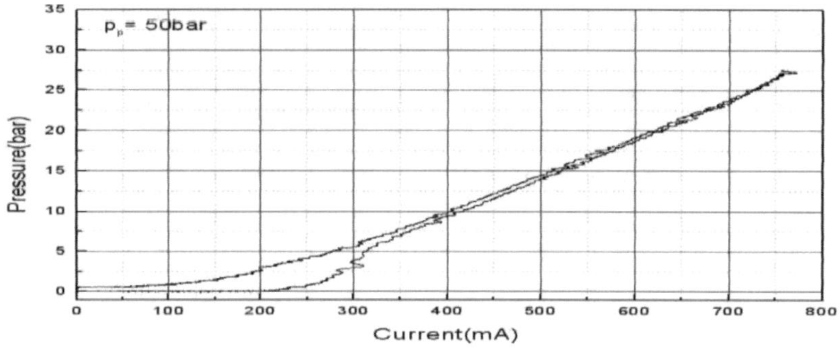

Figure 7. experimental P-I diagram of existent value

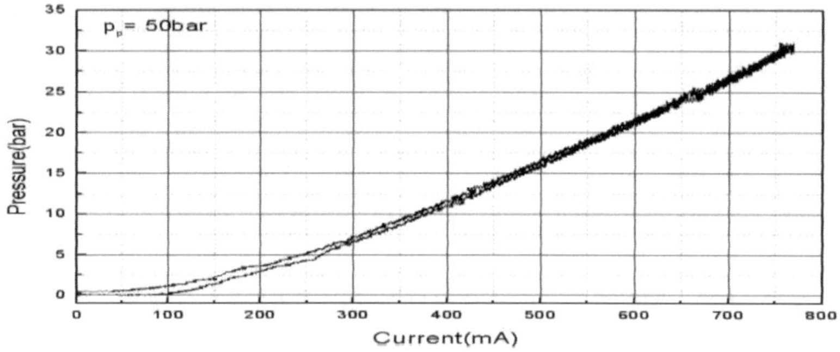

Figure 8. experimental P-I diagram of new value

5. Forklift application experiment of improved EPPR valve

Figure 9. Forlift equipped improved EPPR valve

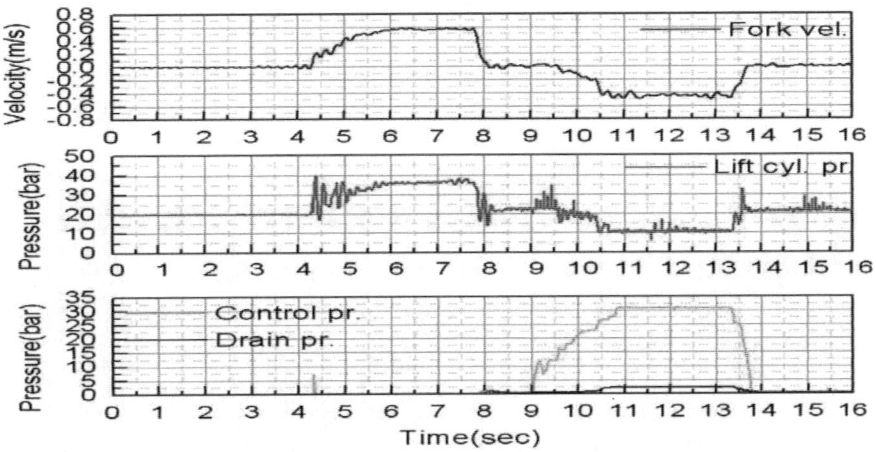

Figure 10. pressure control diagram of improved EPPR valve

6. Results of experimental analysis

The linear pressure range was started from 300mA in existent model as shown in figure 7, but it was started from 100mA in new manufactured model as shown in figure 8. So, The non-linear pressure range of new manufactured model was improved about 30%. The improved linear pressure range is meant that it is able to descend the useful pressure of EPPR valve and the controllable pressure range is extended widely contrast to existent model. Also, the hysteresis was improved properly about the first stage of EPPR operating system. It was shown Forlift equipped improved EPPR valve in figure 9. Though Lift-up valve was operated at 4.3 sec and Lift-down valve was operated from 9 sec to 13.7 sec in figure 10, it was able to know the system was operating very stable.

7. Conclusions

It was proved that the result of simulation using AMESim coincided with result of experiment nearly. And then it was able to suggest the optimum model condition of AMESim program.
It was changed the dimension of EPPR valve parts by the results AMESim simulation and was proved that the sleeve of cone angle, tip width, clearance of plunger and sleeve was very important parameter. The performance of new manufactured model was improved about 30% and the controllable pressure range is extended widely contrast to existent model.
We were able to know that the Forklift operating pressure and velocity was followed suitable and the system was operated very stable in field experiment adapting improved EPPR valve.

Acknowledgments

This paper was supported by Gyeonggi-do and Siheung-city under Industry-University Convergence Campus Outcome Utilization Project, a provincial government financial support program, Korea.

References

[1] Kim B L and Oh, I. K, 2013 *Fall Conference Proceedings KSME* pp 4-5
[2] Kim H M, Tae H J, Lee H W and Lee C H 2003 *Transactions of KSAE* Vol 11 No 6 pp 80-85
[3] Yun S N, Ham Y B and Park P W 2006 *Journal of the KSME (A)* Vol 30 No 10 pp 1202-1208
[4] Tae H J, Jung H I, Kim H M and Lee H W 2003 *Transactions of KSAE Fall Conference* pp 70-76
[5] S N Yun and D Y Yun 2009 *Journal of the KFPS* Vol 6 No 3 pp 28-34
[6] B J Lim and C D Park 2014 *Journal of the KSFC* Vol 11 No 1 pp 26-34
[7] M G Cho and D Y Shin 2010 *Spring Conference Proceedings KSME* pp 1109-1110

Performance Analysis of Hybrid Electric Vehicle over Different Driving Cycles

Aishwarya Panday and Hari Om Bansal

Electrical and Electronics Engineering Department,
Birla Institute of Technology and Science, Pilani, India

hbansal;aishwarya.panday@pilani.bits-pilani.ac.in

Abstract. Article aims to find the nature and response of a hybrid vehicle on various standard driving cycles. Road profile parameters play an important role in determining the fuel efficiency. Typical parameters of road profile can be reduced to a useful smaller set using principal component analysis and independent component analysis. Resultant data set obtained after size reduction may result in more appropriate and important parameter cluster. With reduced parameter set fuel economies over various driving cycles, are ranked using TOPSIS and VIKOR multi-criteria decision making methods. The ranking trend is then compared with the fuel economies achieved after driving the vehicle over respective roads. Control strategy responsible for power split is optimized using genetic algorithm. 1RC battery model and modified SOC estimation method are considered for the simulation and improved results compared with the default are obtained.

1. Introduction

Hybrid electric vehicle (HEV) has turned up as a promising solution for the automotive industries in terms of reduced toxic emissions. As two power sources, battery and engine are present, an intelligent power split energy management strategy should be implemented to split the power to achieve better fuel economy. HEV's performance will obviously vary over the type and conditions of road, driver's aggressiveness/behavior and weather conditions. Kuhlar and Karstens [1] and Fomunung [2] introduced few parameters which characterize the roads. These parameters have been used to model emissions or fuel economies by others. These parameters are able to describe a driving cycle (DC) behavior but can't clearly identify a DC which may result in the best fuel economy (FE). The aim of the study is to find parameters of DCs, which impacts the FE. Using size reduction techniques, few governing parameters are selected and with these parameters DCs are ranked using 'Technique for order preferences by similarity to an ideal solution' (TOPSIS) and 'The Vlse Kriterijum-ska Optimizacija Kompromisno Resenje' (VIKOR). The vehicle is then run over the different considered DCs to find out their FE using genetic algorithm (GA) based control strategy and then ranked accordingly. These two results are compared to get a sense that how the parameters are linked with FE.

In most of the literature ten parameters are generally considered to characterize a DC. These parameters for seven different city DCs are collated in table1. The authors are interested in identifying the dominant parameters. Principle component analysis (PCA) and independent component analysis (ICA) are dimension reducing techniques and retain the important properties of the dataset. PCA developed by Karl [3] and Hotelling [4] is a statistical procedure to un-correlate the variable and reduce the dimension of the data. PCA yields orthogonal vectors of high energy content in terms of covariance. ICA also decomposes the variables into smaller sets and extracts the independent variables from a multivariate dataset. ICA was developed by Jueten and Herault [5] and Comon [6] to solve cocktail party problems and for blind source separation. DC data sets are reduced here using ICA and PCA and some of these reduced parameters are maximized and others are minimized to compute FE. This requires multi-criterion decision making (MCDM) measures. Valuable MCDM methods like TOPSIS and VIKOR, are applied here to rank DCs in the order of their FE. Further, to support the analysis obtained by these methods, an intelligent power split control strategy is developed using GA. Engine on/off trend helps in achieving this. To control the engine on/off, threshold values of governing parameters (which are responsible for power split) are obtained using optimization techniques and FE is determined.

2. Feature extraction of driving cycles

Table 1 lists the parameters of DCs. To extract features of these DCs, PCA and ICA methods are used and explained in the subsequent sections.

Table 1. City driving cycle characteristics

Driving cycle / Parameters	UDDS	ECE_ EUDC	LA92	US-06	Indian	Japan 10-15	WVUCITY
Time (s)	1369	1225	1435	600	2689	660	1408
Distance (miles)	7.45	6.79	9.82	8.01	10.87	2.59	3.3
Max. speed (mph)	56.7	74.56	67.2	80.3	38.87	43.48	35.82
Avg. speed (mph)	19.5	19.95	24.61	47.97	14.54	14.09	8.44
Max. acceleration (ft/s^2)	4.84	3.46	0.12	12.32	5.68	2.6	3.75
Max. deceleration (ft/s^2)	-4.84	-4.56	-12.91	-10.12	-6.9	-2.73	-10.62
Avg. acceleration (ft/s^2)	1.66	1.78	2.21	2.2	1.06	1.87	0.97
Avg. deceleration (ft/s^2)	-1.9	-2.59	-2.47	-2.39	-1.29	-2.12	-1.27
Idle time (s)	259	339	234	45	267	215	427
No of stops	17	13	16	5	52	7	14

2.1. Principal Component Analysis

PCA is useful to the dataset which heavily rely on Gaussian features and utilizes first and second moments of measured data. PCA displays principal components (PCs) which are made of the variables. These PCs are unit vectors and are orthogonal identity matrix. Here, principal component 8 (PC8) onward all PCs are found to be zero. PC1, PC2 and PC3 are significant and are considered for interpretation. PC1 is strongly correlated with eight of the 10 original variables. PC1 can be viewed as a measure of distance, avg. speed, max. acceleration, max. deceleration, avg. acceleration, avg. deceleration and number of stops. DCs with better fuel economy would tend to have lesser values of avg. speed, max. acceleration and avg. acceleration whereas higher values of maximum deceleration, avg. deceleration and number of stops. PC2 increases with only one of the values, i.e. decreasing distance. PC3 increases if the idle time in the driving cycle increases and explains the impact of idle time. Identification of the variables of driving cycle through PCA will be useful in ranking a DC and while analyzing its fuel economy.

2.2. *Independent Component Analysis*

For the hidden feature extraction another popular technique, ICA is also used. ICA is a statistical and computational technique to identify the meaningful hidden features. In contrast to PCA, ICA extracts six parameters which are significant for analysis. These are trip time, average speed, maximum deceleration, average acceleration, idle time during the trip. The smaller data sets received by PCA and ICA do not have same parameters, so selecting one of them is important. Generally, PCA is applicable to Gaussian data and is restricted to the first and second moments of the data, whereas ICA is applicable to non-Gaussian data and explores higher order moments. Thus, it is important to identify the nature of DC data on which these techniques are applied.

3. Nature Identification of driving Cycles

To analyze the nature of DCs, probability distribution functions are determined. The mean and standard deviations (SD) of the distributions are recorded and listed in table 2. It can be observed that no DC follows a normal distribution with mean=0 and SD=1. This enables us to choose ICA for feature extraction as ICA fits for non-Gaussian data set.

Table 2. Mean and standard deviations of various distributions

Driving cycle	Mean (Location parameter)	SD (Scale parameter)	Distribution type
INDIAN	-3.1518	33.790	Generalized Pareto
UDDS	4.2804	5.0278	Gumbel maximum
ECE_EUDC	-1.7264	22.386	Johnson SB
LA92	-1.7373	35.690	Johnson SB
WUVCITY	1.3645	2.4398	Generalized extreme value
US06	-5.1976	80.503	Generalized Pareto
JAPAN10-15	-1.7264	22.386	Johnson SB

4. Ranking method implications

VIKOR and TOPSIS methods are applied to rank the DCs for FE prediction. VIKOR is multi-criteria optimization and a compromise solution method developed by Opricovic and Tzeng [7] and TOPSIS is a multi-attribute decision making method and introduced by Yoon and Hwang [8]. VIKOR and TOPSIS are applied to the extracted features from ICA to rank various DCs in order to get estimation about their FEs without running a vehicle (without simulation). The results for VIKOR and TOPSIS with ICA are tabulated in table 3. The results infer that Indian DC is getting invariably first position with all the ranking methods. Hence, Indian road conditions are proven to be better for HEVs as they will consume lesser fuel and minimum emissions. The low acceleration rate, lower speed and frequent start/stops prompt the motor to work more and keep the engine in shutting off condition or operate in its most efficient region. This becomes a favorable condition for an efficient HEV. Though, other DCs are not found to be of the same rank and a close observation of LA92, UDDS, WVUCITY and ECE_EUDC reveals that most of their parameters are very close to each other, hence taking VIKOR/TOPSIS towards the unintended ranking level.

IOP Conf. Series: Journal of Physics: Conf. Series **806** (2017) 012014 doi:10.1088/1742-6596/806/1/012014

Table 3. Driving cycle ranking by VIKOR and TOPSIS

VIKOR with PCA	VIKOR with ICA	TOPSIS with PCA	TOPSIS with ICA	FE ranking
Indian	Indian	Indian	Indian	Indian
WVUCITY	LA92	WVUCITY	LA92	UDDS
JAPAN 10-15	UDDS	JAPAN 10-15	WVUCITY	ECE_EUDC
UDDS	WVUCITY	UDDS	UDDS	LA92
LA92	ECE_EUDC	ECE_EUDC	ECE_EUDC	WVUCITY
ECE_EUDC	US06	LA92	US06	US06
US06	JAPAN 10-15	US06	JAPAN 10-15	JAPAN 10-15

5. Analysis of fuel economy over different driving cycles

The FE of Prius over considered DCs is measured using GA in ADVISOR. GA is a robust and feasible approach and solves complex optimization problems. The available battery power is governed by state of charge (SOC). ADVISOR uses Ampere Hour Counting method for SOC estimation. But, the importance of open circuit voltage for SOC estimation is also emphasized in literature. Thus, SOC estimation algorithm is accordingly modified and incorporated in ADVISOR library. In this paper, 1RC (parallel combination of 1 RC components in series with internal resistance) model along with modified SOC estimation method is used to perform the simulations. Threshold values of governing parameters responsible to turn on/off the engine are estimated using GA and corresponding FEs are recorded [9]. Table 4 records the threshold values of these parameters, namely cs_eng_on_soc, cs_min_off_time, cs_min_pwr, cs_electric_launch_spd and cs_eng_min_spd for considered DCs. Table 5 collects the efficiencies of engine, motor and generator and FEs over various DCs with the threshold values obtained in table 4. Among all the DCs, Indian urban DC is found to be best in terms of FE. With a tag of the second most populous country today which will become first in 2022 surpassing China, India, will demand deadly number of vehicles on the road to fulfil the requirement. On an average 11-12% per year increase in number of motorized vehicles on Indian roads is observed. India's CO_2 emission continues to increase by 7.8 percent in 2014. India emitted 170 million tons of CO_2 in 2014 and became the largest contributor to global emission growth. India became fourth largest CO_2 emitting country in 2014 while it was fifth in 2011. So, an alternative approach in terms of HEV should be opted to fulfil its growing population demand while reducing the CO_2 emissions. The favorable road conditions paves pathway to invite hybrid vehicles on Indian roads for greener tomorrow.

Table 4. Optimized parameter values of drive cycles using GA

Drive cycle	SOC(%)	Off time(s)	Power(watt)	Electric speed(rad/s)	Enginespeed(rad/s)
UDDS	39.65	9.61	9276.9	10.79	209.43
ECE_EUDC	33.08	6.10	6805.9	12.48	281.66
LA92	38.91	4.26	9306.6	11.08	254.23
US06	38.01	9.80	9318.6	10.63	254.92
Indian	33.60	6.70	9280.2	10.88	252.97
Japan 10-15	33.10	9.71	9762.3	10.64	230.53
WVUCITY	39.11	2.13	9128.3	10.07	230.12

Table 5. Efficiency evaluation of system components using GA

Drive cycle	Engine efficiency(%)	Motor efficiency(%)	Generator efficiency (%)	FE(mpgge)
UDDS	30.30	83.65	64.77	59.32
ECE_EUDC	27.32	81.67	92.46	52.25
LA92	31.48	82.59	67.22	48.18
US06	33.08	82.54	60.58	44.72
Indian	30.34	84.17	23.56	61.44
Japan 10-15	19.16	92.89	80.54	43.01
WVUCITY	21.07	79.39	74.63	46.99

Table 6 performs the comparison between default SOC estimation method with R_{int} battery model and modified SOC estimation method with 1RC battery model. Modified SOC estimation algorithm in table 6 is either improving FE or speed traces or both. In general cases, idling is allowed at the stops. But if, idle stopping is implemented, the significant amount of fuel can be saved. The comparative study of the FEs in case of idle stopping with zero engine speed and engine on is tabulated here. Table 7 shows noteworthy improvement in FEs for all DCs but Indian DC is dominating. Two major reasons idle time and number of stops may play momentous functionality. Idle time is considerable, but a greater no. of stops will recuperate more kinetic energy which will motivate the vehicle to run more on electrical energy, hence improves FE. Further, lower acceleration rate and lower speed of Indian DC make it in favor of HEVs. Indian DC also provides highest regenerative efficiency.

Table 6. Fuel economy comparison with default and modified SOC estimation method

Drive cycle	Fuel economy with default SOC estimation method and R_{int} model	Fuel economy with modified SOC estimation method and 1RC model
UDDS	59 (with trace miss)	59.3291 (without trace miss)
ECE_EUDC	46.7 (without trace miss)	52.2579 (without trace miss)
LA92	44.3 (with trace miss)	48.1884 (without trace miss)
US06	42 (with trace miss)	44.7263 (without trace miss)
Indian	54.7 (with trace miss)	61.4473 (without trace miss)
Japan 10-15	41 (without trace miss)	43.00 (without trace miss)
WVUCITY	46.7 (with trace miss)	46.9988 (without trace miss)

Table 7. Fuel economy comparison with and without engine idling

Drive cycle	FE without EI (mpgge)	FE with EI (mpgge)	FE improve-ment (%)	Idling time (%)	No. of stops	Regenerative efficiency
UDDS	59.32	62.33	5.06	18.91	17	119.591
ECE_EUDC	52.25	54.05	3.44	27.67	13	-78.707
LA92	48.18	50.44	4.69	16.30	16	185.333
US06	44.72	51.58	15.3	7.500	5	217.773
Indian	61.44	85.32	38.8	9.929	52	250.308
Japan 10-15	43.00	44.45	3.37	32.57	7	-381.935
WVUCITY	46.99	48.74	4.61	30.32	14	-193.287

6. Validation of control strategy

The suitability of the proposed control strategy is calculated through numerous simulations performed over different DCs repeated over diverse geographic provinces. The simulations are performed on 1RC battery model with the modified SOC estimation method. At all instances, the demanded power and speeds are met and can be observed in figure 1. Various studies also show that HEVs can give better FE on Indian city roads in comparison to the developed country road [10].

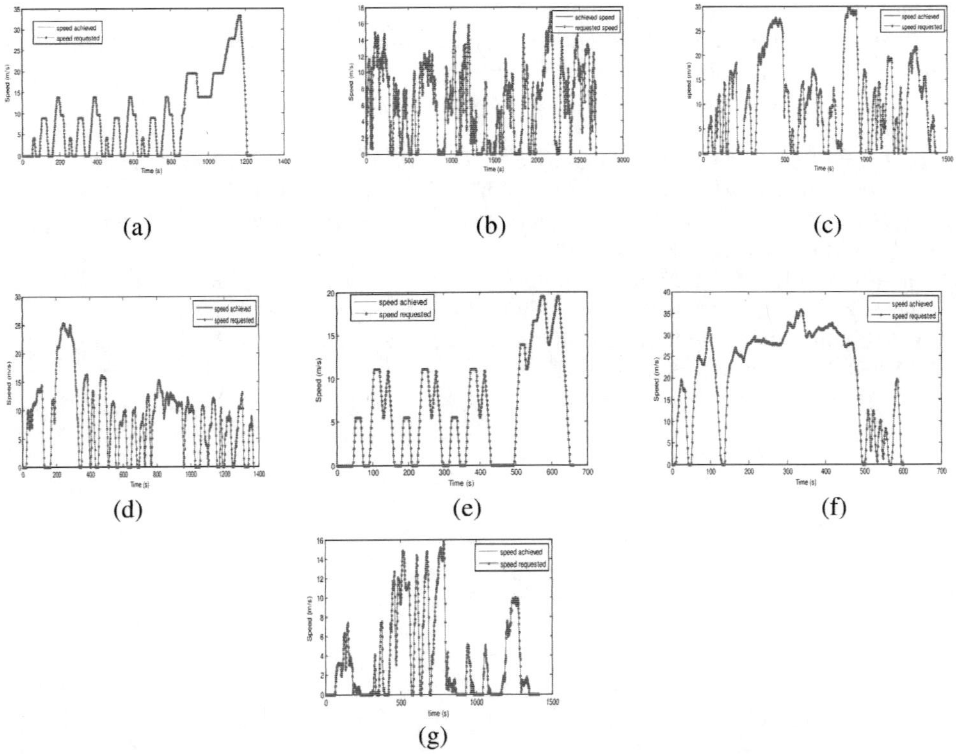

Figure 1. Speed and time traces of driving cycles (a) ECE_EUDC, (b) INDIAN, (c) LA92, (d) UDDS, (e) US06, (f) JAPAN 10-15, (g) WVUCITY

7. Conclusion

The efficiency of an HEV obviously depends on the road profiles. A profile is a composition of various parameters. In this paper, few vital parameters are identified using size reduction techniques and based on these DC are ranked in order of their fuel economy using multi criterion optimization methods. The results are further validated using GA based intelligent power split control strategy. It is concluded that the Indian urban DC is promising and provides higher fuel efficiency for an HEV as compared to other countries. The favorable Indian road profile will attract more people to use HEVs, thus boom in the automobile manufacturing market is expected. This will further reduce toxic emissions and will contribute to the Indian economy. It is also recommended that ICA should be applied rather than PCA for extracting DC parameters to analyze the performance as they follow non-Gaussian distribution. Engine idling should be considered as a powerful feature for improving fuel economy on city roads.

References

[1] Kuhler M and Karstens D 1978 *Improved Driving Cycle for Testing Automotive Exhaust Emissions* (SAE Technical Paper Series 780650)

[2] Fomunung I, Washington S and Guensler R 1999 *A statistical model for estimating oxides of nitrogen emissions from light duty motor vehicles (Transportation Research Part D)* p 333

[3] Pearson K 1901 *On Lines and Phase of Closest Fit to Systems of Points in Space (*Philosophical magazine, vol. 2 (11)) p 559

[4] Hotelling H 1933 *Analysis of a Complex of Statistical Variables into Principal Components* (Journal of Educational Psychology, vol. 24) p 417
[5] Jutten C and Hérault J 1991 *Blind Separation of Sources, Part I: An Adaptive Algorithm Based on Neuromimetic Architecture* (Signal Processing, vol. 24) p 1
[6] Comon P 1994 *Independent Component Analysis-A New Concept*(Signal Processing, vol. 36)p 287
[7] Tzeng G H, Teng M H, Chen J J and Opricovic S 2002 *Multicriteria Selection for a Restaurant Location in Taipei* (*International Journal of Hospitality Management*, vol. 21) p 171
[8] Yoon K and Hwang C L 1981 *Multiple Attribute Decision Making Methods and Applications, A State of the Art Survey* (Springer Verlag, Berlin)
[9] Panday A. and Bansal H O 2016 *Energy Management Strategy for Hybrid Electric Vehicles using Genetic Algorithm* (Journal of Renewable and Sustainable Energy, vol. 8, 015701)
[10] http://www.thehindu.com/news/international/world/untapped-potential-of-hybrid-cars-in-india/article5960835.ece (accessed on 22/01/2016)

A New Supercapacitor and Li-ion Battery Hybrid System for Electric Vehicle in ADVISOR

Xiao Peng[1,2], Quan Shuhai[1,2] and Xie Changjun[2]

[1]School of Automotive engineering, Wuhan University of Technology, Wuhan, 430070, China;
[2]School of Automation, Wuhan University of Technology, Wuhan, 430070, China

xp@whut.edu.cn

Abstract. The supercapacitor (SC) and Li-ion battery(BT) hybrid energy storage system(HESS) electric vehicle(EV) is gaining universal attention. The topology is of importance for the SC/BT HESS.A new SC/BT topology HESS with a rule-based energy management strategy for EV was proposed. The BT pack is connected directly to the DC link via a controlled switch. The SC pack is connected to the DC link via a controlled switch. A uni-directional DC/DC converter is connected between the SC pack and the BT pack. The braking regeneration energy is all harvested by the SC pack.The output power of BT pack is limited. The different SC/BT configurations with varied BT maximum Ah capacity factor and SC maximum capacity factor are simulated in ADVISOR.Simulation results show that BT maximum Ah capacity factor has little impact on vehicle acceleration performance and maximum speed. SC maximum capacity factor has significant impact on vehicle acceleration performance and maximum speed.The fuel economy isn't affected.

1. Introduction

Due to clean and efficient power generation, EV have received increased attention[1].At the same time, challenges such as vehicle performance and cost of EV are yet to be overcome. Li-ion BT is with high energy density and low power density[2,3,4,5].SC is well known for its extremely high power density, high cycle lifetime, cycling efficiency and low energy density[6,7,8]. A BT/SC HESS EV is a possible solution to enhance the energy density and power density at the same time.

The topology of BT/SC HESS is important. A SC/BT topology with a bi-directional DC-DC converter is studied[9,10,11,12,13].A BT/SC topology is developed with a bi-directional converter[14].A BT/SC topology for tramway with two converters is adopted[15].

2. A New SC/BT HESS topology

A proposed new SC/BT HESS topology with two controlled switches is proposed as figure 1. The features are that a uni-directional DC/DC converter is used and all the braking regeneration energy is harvested by the SC pack .The controlled switch 1 is off all the time to protect BT pack isolated from charging current. The controlled switch 2 is on when EV is in braking regeneration phase.The controlled switch 2 is off when the EV is in traction phase. The voltage of SC pack is lower than that of BT pack. The voltage of SC pack is boosted by the DC/DC converter to supply power.

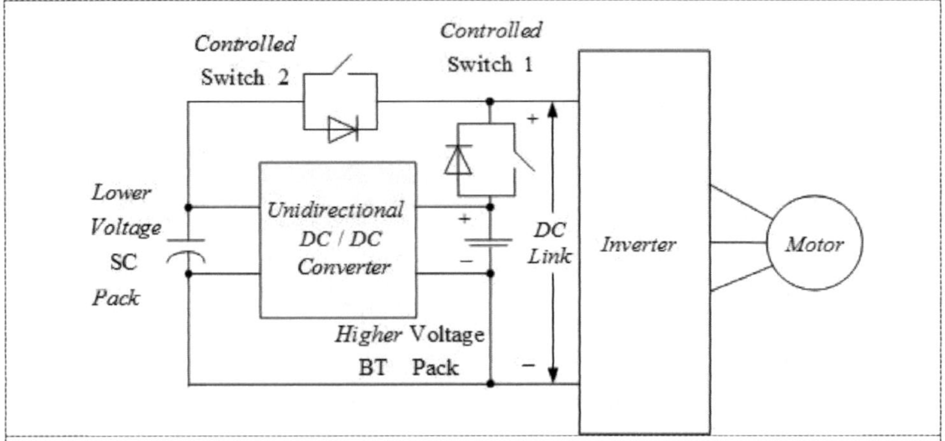

Figure 1. A proposed new SC/BT HESS topology.

3. Rule-based energy management strategy

The power allocation between BT and SC is of importance in the HESS.The power allocation scheme is shown in figure 2.

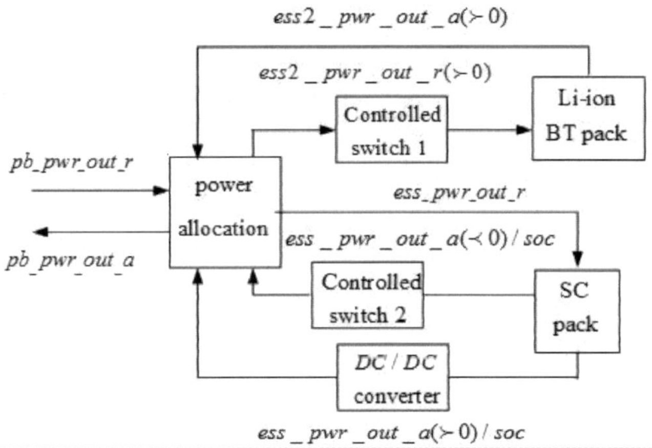

Figure 2. HESS power allocation scheme in ADVISOR.

$pb\_pwr\_out\_r$ is power requested for EV. $pb\_pwr\_out\_a$ is HESS output power for EV. $ess2\_pwr\_out\_r$ is power requested from Li-ion BT pack. $ess2\_pwr\_out\_a$ is output power from Li-ion BT pack. $ess\_pwr\_out\_r$ is power requested from SC pack. $ess\_pwr\_out\_a$ is output power from SC pack. η_{DC} is the constant efficiency of the DC/DC converter. The power allocation is as (1),(2).

$$pb\_pwr\_out\_r = ess2\_pwr\_out\_r + ess\_pwr\_out\_r \quad (1)$$

$$pb\_pwr\_out\_a = ess2\_pwr\_out\_a \cdot \eta_{DC} + ess\_pwr\_out\_a \quad (2)$$

The rule-based energy management strategy is as:

(1) $ess2\_pwr\_out\_a$ is limited between the minimum operating power $cs\_\min\_pwr$ and the maximum operating power $cs\_\max\_pwr$.

(2)When EV is in traction phase, $ess2\_pwr\_out\_a$ is as (3).
(3) When EV is in braking regeneration phase, SC pack receives all braking regenerative power.
(4)The SC state of charge (SOC) is maintained near the goal SOC $soc_{\_goal}$ (As (4)).

$$ess2\_pwr\_out\_a = pb\_pwr\_out\_r + pwr_{\_additional} \quad (3)$$

Where $pwr_{\_additional}$ (as(5)) is additional power to maintain the SC SOC in equilibrium.

$$soc_{\_goal} = 0.5 \cdot (cs\_hi\_soc + cs\_lo\_soc) \quad (4)$$

$$pwr_{\_additional} = \frac{soc_{\_goal} - SOC}{0.5 \cdot (cs\_hi\_soc - cs\_lo\_soc)} \cdot cs\_charge\_pwr \quad (5)$$

Where cs_hi_soc is the SC SOC high limit. cs_lo_soc is the SC SOC low limit. cs_charge_pwr is charge power.

4. Simulation and results

A Saft VL45E LiFePO4 BT [12] and a Maxwell BACP3000 SC are used .The SC/BT HESS EV is simulated based on secondary development in ADVIOSR . Vehicle components parameters for simulation is as table 1. veh_cargo_mass represents cargo mass. veh_glider_mass represents vehicle mass without powertrain. ess2_parallel_mod_num represents the number of BT in parallel.

Table 1. Vehicle components parameters for simulation.

Component	Parameter
Vehicle	veh_cargo_mass 136kg veh_glider_mass 592.4kg
Li-ion BT	ess2_parallel_mod_num 1
SC	ess_module_num 130
Powertrain Control	cs_hi_soc 0.9 cs_min_pwr 1500W cs_lo_soc 0.7 cs_max_pwr 9500W
Motor	continuous power 30kW,peak power 45kW
DC/DC converter	efficiency 0.98

The configurations with varied $ess2\_cap\_scale$, ess_cap_scale, ess_parallel_mod_num are used to observer the vehicle performance when ess_module_num is 130. Vehicle performance ($ess2\_cap\_scale$ = 1, ess_cap_scale = 0.5) is shown in table 2.Vehicle performance ($ess2\_cap\_scale$ =2, ess_cap_scale =0.5) is shown in table 3. Vehicle performance ($ess2\_cap\_scale$ =1, ess_cap_scale =1) is shown in table 4. Vehicle performance ($ess2\_cap\_scale$ =2, ess_cap_scale =1) is shown in table 5. ess_parallel_mod_num is the number of SC string in parallel. ess_module_num is the SC number in series. ess2_module_num is the BT number in series. $ess2\_cap\_scale$ is the factors of BT max. capacity. ess_cap_scale is the factors of SC max. Capacity.

Table 2. Vehicle performance ($ess2\_cap\_scale$ =1, ess_cap_scale = 0.5).

BT Units	SC Units	Fuel Economy (mpgge)	Acceleration(s) 0-60(mph)	Max. Speed (mph)
ess2_module_num=110	ess_parallel_mod_num=1	128.7	N/A	44.9
ess2_module_num=110	ess_parallel_mod_num=2	129.3	10.7	63.7
ess2_module_num=110	ess_parallel_mod_num=3	129.3	8.8	74.7
ess2_module_num=110	ess_parallel_mod_num=4	129.3	8.3	82

Table 3. Vehicle performance ($ess2\_cap\_scale$ =2, ess_cap_scale = 0.5).

BT Units	SC Units	Fuel Economy (mpgge)	Acceleration(s) 0-60(mph)	Max. Speed (mph)
ess2_module_num=110	ess_parallel_mod_num=1	128.9	N/A	43.9
ess2_module_num=110	ess_parallel_mod_num=2	128.9	10.9	64.6
ess2_module_num=110	ess_parallel_mod_num=3	128.8	8.9	72.7
ess2_module_num=110	ess_parallel_mod_num=4	125.5	8.2	81.3

Table 4. Vehicle performance ($ess2\_cap\_scale$ =1, ess_cap_scale = 1).

BT Units	SC Units	Fuel Economy (mpgge)	Acceleration(s) 0-60(mph)	Max. Speed (mph)
ess2_module_num=110	ess_parallel_mod_num=1	132.2	N/A	56.4
ess2_module_num=110	ess_parallel_mod_num=2	132.2	9.4	68.9
ess2_module_num=110	ess_parallel_mod_num=3	131.1	8.4	78.8
ess2_module_num=110	ess_parallel_mod_num=4	132	8.1	85.5

Table 5. Vehicle performance ($ess2\_cap\_scale$ =2, ess_cap_scale = 1).

BT Units	SC Units	Fuel Economy (mpgge)	Acceleration(s) 0-60(mph)	Max. Speed (mph)
ess2_module_num=110	ess_parallel_mod_num=1	132.1	N/A	56.1
ess2_module_num=110	ess_parallel_mod_num=2	132.1	9.7	68.1
ess2_module_num=110	ess_parallel_mod_num=3	131	8.6	78.1
ess2_module_num=110	ess_parallel_mod_num=4	130.3	8.3	84.6

It can be seen in table 2 and table 3,when ess_cap_scale =0.5, $ess2\_cap\_scale$ is increased from 1 to 2, the vehicle acceleration performance is decreased by 1.1%-1.8% and max. speed is decreased by 0.86%-2.2%.

It can be seen in table 4 and table 5,when ess_cap_scale =1, $ess2\_cap\_scale$ is increased from 1 to 2, the vehicle acceleration performance is decreased by 2%-3% and max. speed is decreased by 0.5%-1.2%.

It can be seen in table 2 and table 4,when $ess2\_cap\_scale$ =1, ess_cap_scale is increased from 0.5 to 1,the vehicle acceleration performance is improved by 2.4%-13.8% and max. speed is improved by 4.2%-25.6%.

It can be seen in table 3 and table 5,when $ess2\_cap\_scale$ =2, ess_cap_scale is increased from 0.5 to 1,the vehicle acceleration performance is improved by 1.2%-12.3% and max. speed is improved by 4%-27%.

The fuel economy isn't affected.

5. Conclusion

A new SC/BT topology HESS with a rule-based energy management strategy for EV was proposed. The braking regeneration energy is all harvested by the SC pack. A uni-directional DC/DC converter is used.EV performance are studied . BT maximum Ah capacity factor has little impact on vehicle acceleration performance and maximum speed. SC maximum capacity factor has significant impact on vehicle acceleration performance and maximum speed.The change of fuel economy isn't obvious.The conclusion is beneficial to the EV sizing and optimization.

Acknowledgement

This research was supported by the National Natural Science Foundation of China (51477125), and th e national key basic research development plan (973 Plan) (No. 2013CB632505), the Hubei province t echnological innovation major project (2016AAA046), and the Wuhan youth morning project (201607 0204010155).

References

[1] Troy R. Hawkins, Ola Moa Gausen, Anders Hammer Strømman,"*Environmental impacts of hybrid and electric vehicles—a review,*". International Jouanalof Life Cycle Assessment,vol.17,no.8,pp.997–1014,2012.

[2] Scott B. Peterson, Jay Apt , J.F. Whitacre. *Lithium-ion battery cell degradation resulting from realistic vehicle and vehicle-to-grid utilization.* Journal of Power Sources 195 (2010) 2385–2392.

[3] Bruno Scrosati, Jürgen Garche.Lithium batteries: *Status, prospects and future.*Journal of Power Sources 195 (2010) 2419–2430.

[4] WladislawWaag , Christian Fleischer , Dirk Uwe Sauer .*Critical review of the methods for monitoring of lithium-ion batteries in electric and hybrid vehicles.*Journal of Power Sources 258 (2014) 321-339.

[5] J. Vetter , P. Novak´ , M.R. Wagner , C. Veit , K.-C. Moller etc.*Ageing mechanisms in lithium-ion batteries.*Journal of Power Sources 147 (2005) 269–281.

[6] Xuebing Han, Minggao Ouyang, Languang Lu, Jianqiu Li, Yuejiu Zheng, ZheLi.*A comparative study of commercial lithium ion battery cycle life in electrical vehicle: Aging mechanism identification.*Journal of Power Sources 251 (2014) 38-54.

[7] Languang Lu, Xuebing Han, JianqiuLi,"*A review on the key issues for lithium-ion battery management in electric vehicles,* "Journal of Power Sources,vol.226 ,no.3,pp. 272-288,2013.

[8] Marina Mastragostino , Francesca Soavi.*Strategies for high-performance supercapacitors for HEV.*Journal of Power Sources 174 (2007) 89–93.

[9] Changle Xiang , Yanzi Wang , Sideng Hu and Weida Wang .*A New Topology and Control Strategy for a Hybrid Battery-Ultracapacitor Energy Storage System.*Energies 2014, 7, 2874-2896

[10] Micah Ortuzar,JorgeMoreno,Juan Dixon," *Ultracapacitor-Based Auxiliary Energy System for an Electric Vehicle: Implementation and Evaluation,*" IEEE Transactions on industrial electronics,vol. 54, no. 4, pp.2147-2156,2007.

[11] Ali Castaings, Walter Lhomme, RochdiTrigui, Alain Bouscayrol.*Practical control schemes of a battery/ supercapacitor system for electric vehicle.*Special Issue: Energy Storage and Electric Power Sub-Systems for Advanced Vehicles (Based on IEEE-VPPC 2014).2016, Vol. 6, Iss. 1, pp. 20–26.

[12] Junyi Shen, SerkanDusmez,AlirezaKhaligh,"*Optimization of Sizing and Battery Cycle Lifein Battery/Ultracapacitor Hybrid Energy Storage Systems for Electric Vehicle Applications,*"IEEE Transactions on industrial informatics,vol.10,no.4,pp.2112-2121,2014.

[13] Victor Isaac Herrera, HaizeaGaztañaga, Aitor Milo, AndoniSaez-de-Ibarra, Ion Etxeberria-Otadui.*Optimal Energy Management and Sizing of a Battery - Supercapacitor based Light Rail Vehicle with Multi - objective approach.*IEEE Transactions on Industry Applications.Citation information: DOI 10.1109/TIA.2016.2555790.

[14] Jian Cao,AliEmadi,"*A New Battery UltraCapacitor Hybrid Energy Storage System for Electric,Hybrid,and Plug-In Hybrid Electric Vehicles,*"IEEE Transactions on power electronics,vol. 27,no.1,pp.122-132,2012.

[15] Victor Herrera,AitorMilo,etc.*Adaptive energy management strategy and optimal sizing applied on a battery-supercapacitor based tramway.*Applied Energy 169（2016）831-845.

Association Sorting Algorithm Design for Error Searching System

Fan Yang, Zhenghong Dong and Mengwei Li

Academy of Equipment, Beijing, China

sarah0824@hotmail.com

Abstract. For the searching results of error searching system for Integrated Decision Information System (IDIS), this paper proposed a method to carry out association sorting according to degree of association of searched results by keywords, which can priorily recommend user interested searching results. Errors of IDIS platform are occurred very often. Because those errors belong to different stages like setup, configuration, and operation, or those errors may occurred in different services, applications, or IP ports, or may be happened in different system software, different version of software, and those errors are also can be classified into different types. As there will be many error information are searched out, it is required to sort them according to degree of association, and provide user interested results. This paper proposed a method to carry out association sorting for the searching results according to the times of searched repeated results by keywords, which effectively put results with high degree of association in the front and increase the searching efficiency.

1. Introduction

IDIS integrates a lot of sub-system from different institutes, including almost a hundred different software. Because of the differences of underlayer protocol, and the difference of standard, there are many errors occurred during the stages of setup, configuration, and operation, which seriously affects the usage. Moreover, because the errors are various, which may happen in different operation phases, stages, TCP/IP communication protocol layers, sub-system software, it is necessary to design a database system which can manage those errors. The proposed method provide a design of error searching database, which can make the correspondence among error, the reason of the error, and corresponding solution, and put them to different categories in terms of their characteristics, such that it is easy to manage, search, and use. Users do not only search the reason and solution of an error, but also can find out the accurate information of the error, such as in which stage, layer, software, or categories, etc. the error has happened.

The current error searching systems are various, including On-Board Diagnostics designed for vehicles [1], which can search vehicle errors according to vehicle OBD error code; Computer error searching system [2], which can search the hardware problems, network problems, and software problems, etc. Those error searching systems focus on errors in specific area, design different databases and the associations in between to store the errors as well as the characteristics of those errors. Because IDIS is used for communication and ordering between the superiors and the subordinates in specialized field, the design for DB tables, logical structures has to build up according the specialized characters of IDIS, and the design and definition for keywords requires personalized customization. In that case, the current error searching software cannot be applied to this platform. It is necessary to design a specialized database system to effectively solve the various errors occurred in the IDIS.

For the searching results of error searching system for IDIS, this paper proposed a method to carry out association sorting for the searching results according to the times of searched repeated results by keywords, which effectively put results with high degree of association in the front and increase the searching efficiency.

For the arrangement of this paper, section 2 will discuss related work, analysis their advantages and disadvantages. Section 3 will introduce the proposed method, including the design on the system function, as well as the flowchart of this system. Section 4 provides the system prototype and section 5 makes the conclusion of this paper.

2. Related Work

The current sematic searching algorithms only focus on the association between information about the searching results. They only focus on the structure of information, which leads that user expected searching results have been put to the back of the results. Reference [3] has proposed a sorting algorithm based on genetic algorithm, which optimizes the sorting results.

A lot of work [4-8] focused on analyzing the data properties, such as the usage of tagging patterns, the semantics of tags, the usage of annotations, and other personalized user behavior. Reference [9] combines the contents of web pages and the social annotations as well as user's personalized preferences to search the result, which can be applied to social annotation information system. Reference [10] ranks search result based on user-label, which requires user feedbacks and user queries. However, this method requires large amount of data to make the ranking results reliable, otherwise, the ranking results are invalid if the sample data is limited. Apart from the content of search results, reference [11] also considers time dimension for ranking. Because people may not only considered about the relevance, but also would like to view the latest article. In this reference, new articles would be recommended to users. However, it is hard to define how well the time dimension would affect the ranking results.

Sematic relevance is well considered in current works for ranking search results, which is quite useful to find the relevance between user query and search results. This is the most important factor that controls the ranks of search results. So this paper focuses on sematic relevance to rank search results, which considered about the quality and quantity of user queried keywords that matches with the search results. And this way of rank is more reliable.

3. Proposed Method

3.1. System function design

Figure 1 shows the system function design. The error database system contains two modules, which are search engine and database, where database has three sub-function modules (D1, D2, and D3).

D1: Data import/enter

This function supports two ways of importing data. One is to import by Excel file directly, and the other is to enter data by administrator.

D2: Keyword fuzzy search

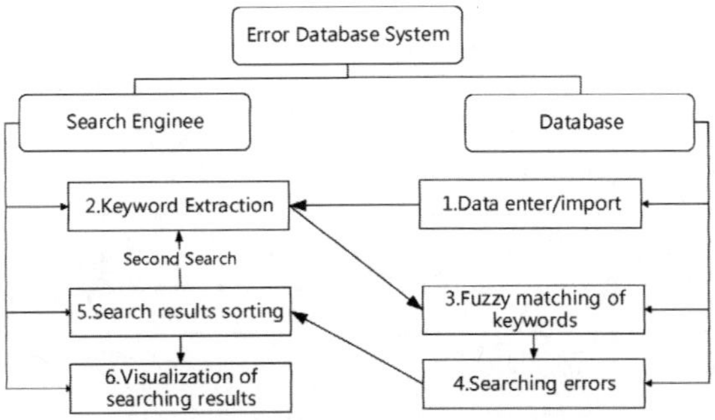

Figure.1 System function diagram.

This function supports for the fuzzy matching between the extracted keywords and the keywords in keyword table.

D3: Find out the corresponding error, reason, and solution according to the extracted keywords.

Search engine includes 4 sub-function modules (S1, S2, S3, and S4).

S1: Keyword extraction

To extract useful keywords from the input contents.

S2: Association Sorting for the searching results

Association Sorting Algorithm: to sort the searching results according to matching degree.

Degree of association sorting algorithm has been proposed to priorly recommend more matched results.

Considering that the researched error information contains multiple records, the proposed method priorly recommends more matched searching results according to sorting results based on the degree of association of keywords

First, to consider the errors information that contains filtering conditions (error characters), and put those errors to the first half of the searching results list. While for the searching results that are searched by keywords with priority equals to 2, we put them to the latter half of the searching results list.

Next, to sort the first half and the latter half of results list, respectively. According to the keywords to find out corresponding error ID from the table of keyword and error relation. Because an error has several keywords, so we can find out the same error ID many times according to those keywords. If an error ID has been searched out more times, then it means it contains more keywords, and it is more related to the searched error. Then we sort the searching results according to errors times that have been searched out, so that user interested results can be put to the front.

S3: Second search

The system also supports second search from the already searched results.

S4: Visualization of searching results

This is to display the search results visually.

3.2. Flowchart

Figure 2 gives the whole flow chart of the error searching system which contains 5 stages, which are keyword extraction, fuzzy matching, find out error ID, sorting search results, and visualization of searching results, respectively. As sorting process only evolves stage 4 and stage 5, so we explain the 2 stages in more detail as table 1.

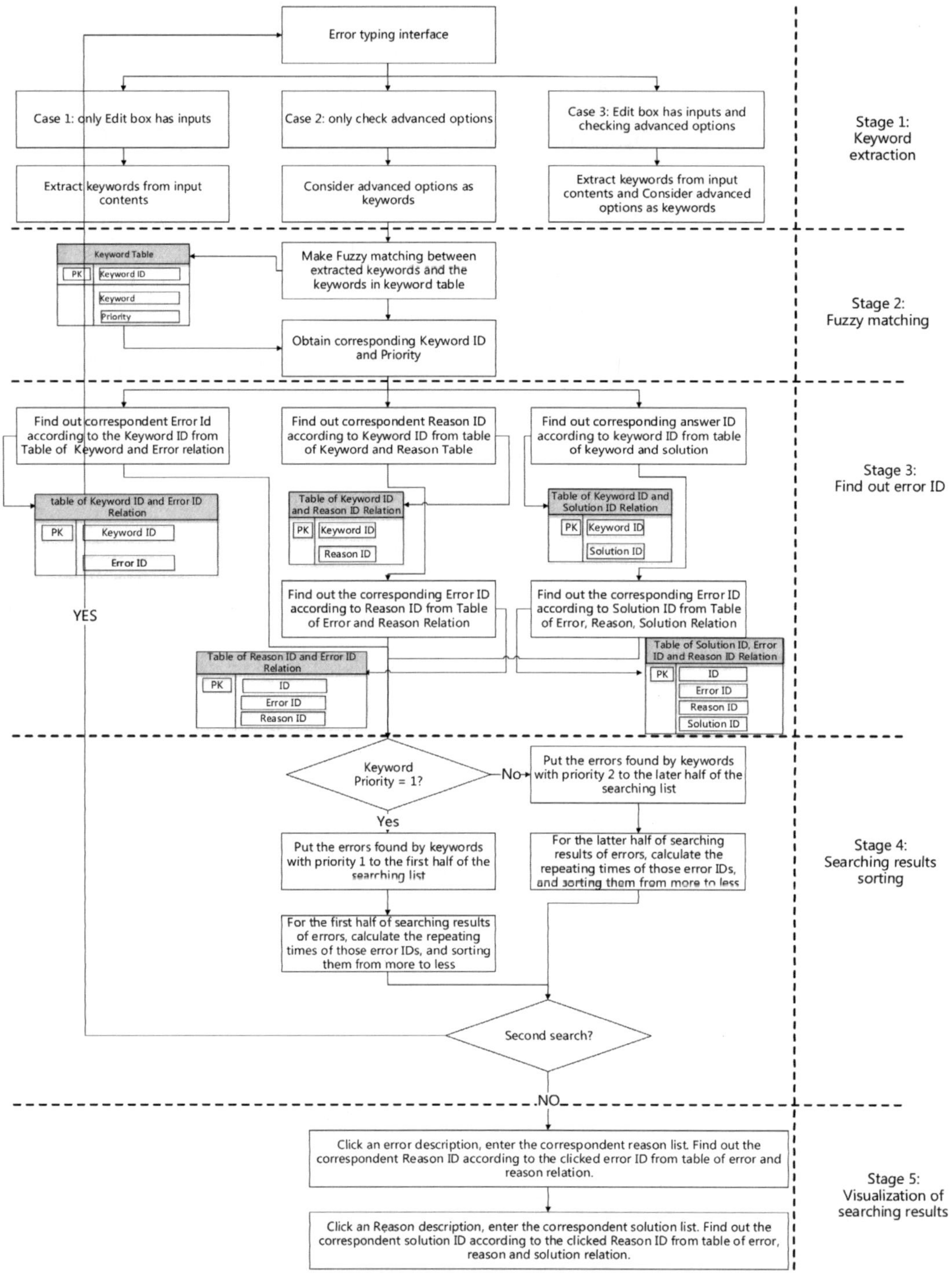

Figure 2 Flow chart of the error searching system.

Table. 1 Explanation on Database searching flowchart

Searching stages		**Contents**
Stage 4 Sorting up searching results	1.	Find out a set of Errors by keyword with priority=1, and put them to the first half of searching result list. To compute the times that have been found out, and sort the set of errors according to the searched times from more to less, which is the first half of searching result list.
	2.	Find out a set of Errors by keyword with priority=2, and put them to the latter half of searching result list. To compute the times that have been found out, and sort the set of errors according to the searched times from more to less, which is the latter half of searching result list.
	3.	If there is no such result has been found according keyword database, then directly carry out fuzzy matching with the errors in the table of error.
	4.	If user would like to narrow the searching scale, he can apply second search, the same searching method applies here.
Stage 5 Searching results visualization	1.	After error list has been generated in terms of degree of association from high to low, according to error ID, to find out corresponding reasons from the table of error and reason relation, and generate reason list. The way of presentation is like this, click on an error description, the correspondent reason list is displayed.
	2.	After reason list has been given, according to error ID and reason ID, to find out corresponding solutions from the table of error, reason, and solution relation, and generate solution list. The way of presentation is like this, click

4. Experiments and Prototype

This paper has implemented a prototype of this searching system for verifying if the sorting results are valid and useful. We have shown the prototype by figure 3. Figure 3 shows the searching results sorting page. After search for keyword "system", the searching results are list as figure 4. We can see from the figure, related items that relevant to "system" are put to the front of research results. And even if one item does not contain the exact word of "system", it still could be put to the front if the reason or solution of this question contains the word, shown as figure 5.

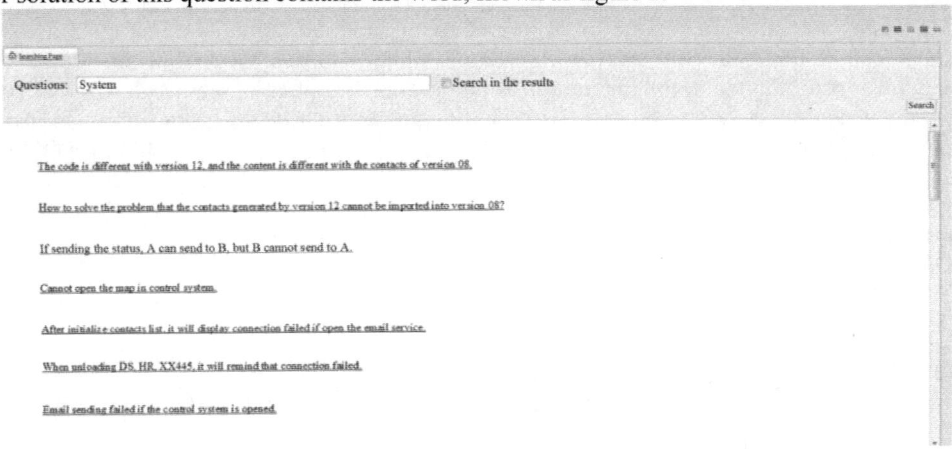

Figure 3 Search results sorting page

IOP Conf. Series: Journal of Physics: Conf. Series **806** (2017) 012016 doi:10.1088/1742-6596/806/1/012016

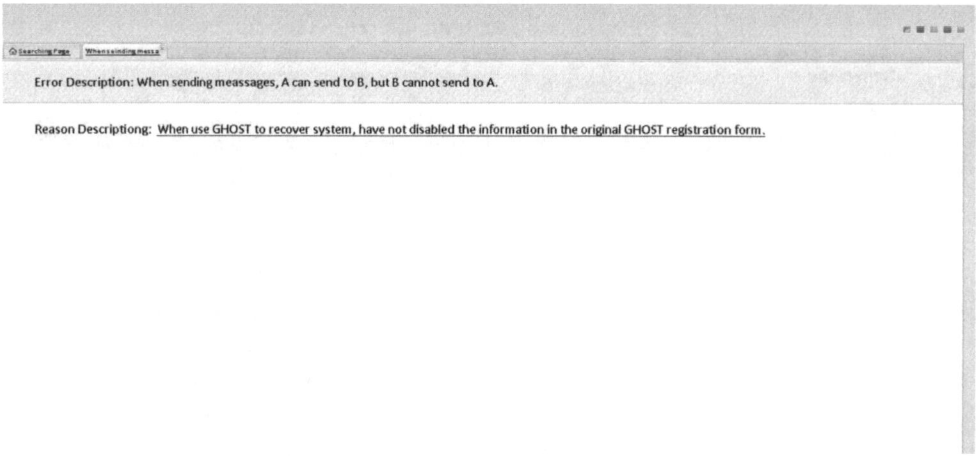

Figure 4 Page of reason that contains the word "system".

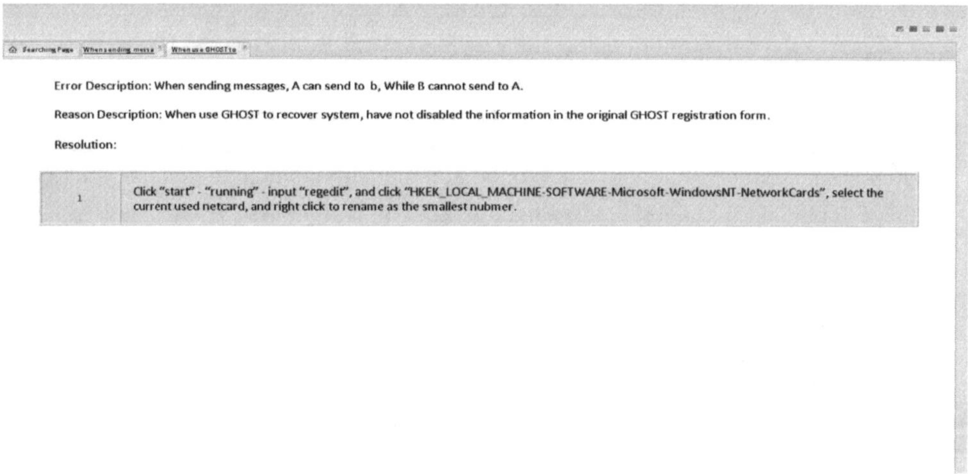

Figure 5. Page of solution that contains the word "system"

5. Conclusion

For the searching results of error searching system for IDIS, this paper proposed a method to carry out association sorting for the searching results according to the times of searched repeated results by keywords, which effectively put results with high degree of association in the front and increase the searching efficiency. We have provided a prototype to show the sorting results and explain the sorting idea.

References

[1] Wang, J.H., Fang, M.D., Gao, J.D., Lu, H.Y. and Dai, C.B. Basic Principle and Application of On-Board Diagnostics for Gasoline Fuelled Vehicles. Automotive Engineering, 28, 491-494, 2006.

[2] Lu, C., Yang, Y.-H. and Xu, G.-M. Exploitation of Computer Problem Repair and Require on Web System, Journal of Zhejiang Ocean University, 2008.

[3] Yahui Wang, Sematic Searching results sorting algorithm based on Genetic

Algorithm, Journal of Jincheng Institute of Technology, 6(1), 2012.

[4] S. Golder and B. A. Huberman. Usage patterns of collaborative tagging systems. Journal of Information Science, pages 198–208, 2006.

[5] X. Wu, L. Zhang, and Y. Yu. Exploring social annotations for the semantic web. In WWW '06: Proceedings of the 15th international conference on World Wide Web, pages 417–426, New York, NY, USA, 2006.

[6] P. A. Dmitriev, N. Eiron, M. Fontoura, and E. Shekita. Using Annotations in Enterprise Search. In Proc. of WWW 2006, pp.811-817, 2006.

[7] S. Xu, S. Bao, B. Fei, Z. Su and Y. Yu. Exploring Folksonomy for Personalized Search. In Proc. Of SIGIR 2008, pp. 155-162, 2008.

[8] M. Noll and C. Meinel. Personalization 2.0: Web Search Personalization Via Social Bookmarking and Tagging. In Proc. Of ISWC 2007 + ASWC 2007, pp. 567-581, 2007.

[9] C-C Zhao, Z-Q Zhang, H-L Li, and X-Q Xie, A search result ranking algorithm based on web pages and tags clustering, IEEE International Conference on Computer Science and Automation Engineering, 4, 2011.

[10] Z-Q Zhang, T-T Liang, X-Q Xie, "An user-label based search result ranking algorithm", Journal of Computer research and development, 46(supply): 351-358, 2009.

[11] Shu-Hua Chen, Han Yan, Jian-Cheng Li, Jun-Jie Wang, Yu-Ting Mao, Jun Liu, "Improvement of PageRank Algorithm: An Authoritative and Temporal Based Approach", IEEE conference on Anthology, 2013.

Single-Pass Clustering Algorithm Based on Storm

LI Fang, DAI Longlong, JIANG Zhiying and LI Shunzi

Beijing University of Chemical Technology, Beijing, 100029

lifang@mail.buct.edu.cn

Abstract. The dramatically increasing volume of data makes the computational complexity of traditional clustering algorithm rise rapidly accordingly, which leads to the longer time. So as to improve the efficiency of the stream data clustering, a distributed real-time clustering algorithm (S-Single-Pass) based on the classic Single-Pass [1] algorithm and Storm [2] computation framework was designed in this paper. By employing this kind of method in the Topic Detection and Tracking (TDT) [3], the real-time performance of topic detection arises effectively. The proposed method splits the clustering process into two parts: one part is to form clusters for the multi-thread parallel clustering, the other part is to merge the generated clusters in the previous process and update the global clusters. Through the experimental results, the conclusion can be drawn that the proposed method have the nearly same clustering accuracy as the traditional Single-Pass algorithm and the clustering accuracy remains steady, computing rate increases linearly when increasing the number of cluster machines and nodes (processing threads).

1.Introduction

With the development of the Internet, accessing information from the network has become a part of daily life. Therefore, the research of Topic Detection and Tracking (TDT) has attracted more and more attentions. TDT is a processing technology what is aiming at unknown information identification and known information tracking. TDT makes the topic occurred in special time and place extend to extension of topic with more relevance, which becomes an important technology for the decision makers to obtain necessary information.

It is well known that this is the era of big data[4], which is characterized by a large amount of data, complex structure, faster generated data, and more data types. Therefore, the traditional Single-Pass algorithm couldn't handle such a large amount of data. Clustering and detecting of large amounts of data become increasingly difficult and timeliness is getting lower and lower. So in this paper, a Single-Pass clustering algorithm based on Storm framework is used to identify and track text data. Storm is based on Hadoop [5] distributed, fault-tolerant, real-time computing systems, and makes up a

Hadoop batch not meet real-time requirements. Mark and Li Lingjuan achieve the CluStream algorithm in Storm[6], improving the speed of the cluster, and the accuracy has not changed. Wang Mingkun implemented DBSCAN algorithm in Storm [7] and increased throughput. All of them are illustrated by Storm framework can be used in clustering. Therefore, this paper trying to achieve combine Single-Pass algorithm with the Storm framework and improved the algorithm. As Single-Pass algorithm for clustering data can be affected by the order of the results, this paper proposed a two-tier clustering algorithm, ensuring the accuracy of the algorithm. Proved by testing new algorithms to ensure the accuracy of cluster at the same time, which greatly improves the speed and improve timeliness.

2.Operating Mechanism

2.1 Development of Storm

With the development of big data, Hadoop, a software framework which can use on big data to reliable, efficient, and can telescopic of way for distributed processing, came into being. The Mapreduce[8] framework can improve the algorithm parallel degrees and operation speed. However, Hadoop can only process the offline data, and real time cannot be guaranteed. From 2011 the Storm open source, its high throughput, high efficiency greatly improved the processing of big data.

2.2 Mechanism of Storm

Storm cluster consists of a master node and many worker nodes. Master node runs the "Nimbus" daemon, also known as Nimbus nodes for distributing code, assigning tasks, and fault detection. Each node runs the "Supervisor" daemon which has become a Supervisor node for listening, starting and terminating the worker process. Between master nodes and worker nodes in the Storm, there is Zookeeper cluster. Supervisor node contains a lot of Worker nodes, specific tasks are performed by the Worker nodes. Storm framework is shown in the following example, 1:

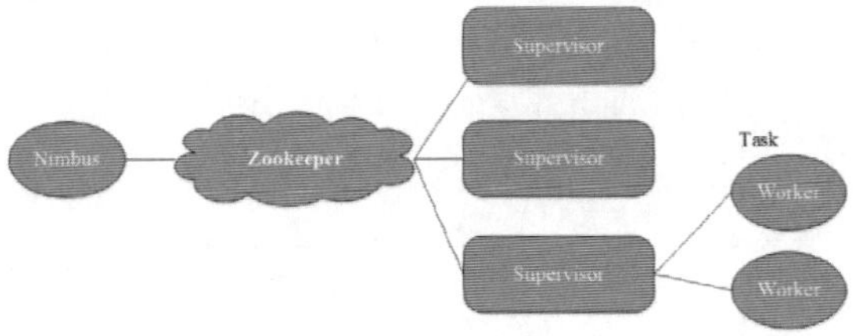

Figure 1. the framework of storm

Topology is the logical units of Storm. Will be packaged into the calculation task after the topology is published, task topology is composed of many of the spout node and bolt node. Spout is the node of reading data, through spout node can read data from external data source (MySQL, and HBase) and to metal group of form sent to corresponding of bolt node, calculation and processing by bolt node, bolt node also can sent the results of their operation to next layer of bolt node, such on put task into multiple part, each part is parallel processing, speed up the whole task. Task topology after submission will run unless end by others, in line with the characteristics of stream data processing.

3.Distributed Real Time Clustering Algorithm For Stream Data

3.1 TF–IDF Algorithm

This paper uses data from Guizhou province Research Center of public opinion text data, stored in a cluster of HBase. Text based on vector space model (VSM) [9], expressed in the form, can have a text representation

$$D=\{(t_1,w_1), (t_2,w_2), (t_3,w_3)\ldots (t_n,w_n),\}$$

The t_1, t_2, t_3 ... t_n is feature for text content, And the w_1, w_2, w_3, w_n is corresponding feature weights, Then the weight calculated by TF–IDF algorithm.

Term frequency–inverse document frequency (TF–IDF) is one of the most fundamental techniques in IR which is a product of term frequency (TF) and inverse document frequency (IDF) where TF represents the importance of the term in a document and IDF represents the importance of term in entire corpus. Equation (1) (2) is as follows:

$$tf_{i,j} = \frac{n_{i,j}}{\sum_k n_{k,j}} \tag{1}$$

$$idf_i = \log \frac{|D|}{|\{j : t_i \in d_j\}|} \tag{2}$$

The title and content is usually divided into two parts, the title usually as the content of the summary expression, more important than content, and the weights for title should be increased, so this paper gives the weights a coefficient (coefficient >1).

3.2 Computes the Degree of Similarity between Two Documents

After vectorization of text, we usually compute the similarity of the text and compare the distance with the threshold size for text clustering. Text vector will be sent Shuffle grouping from this bolt node to the next node. On the second layer of bolt node, text space distance between the texts which usually is to calculate the cosine of the angle between two text vectors. Equation (3) is as follows:

$$\cos\theta = \frac{\sum_{i=1}^{n}(A_i \times B_i)}{\sqrt{\sum_{i=1}^{n} A_i^2} \times \sqrt{\sum_{i=1}^{n} B_i^2}} \tag{3}$$

When the cosine values closer to 1, show the closer of the two texts, that is, the more likely of the two texts belong to the same cluster. Due to calculation text each other similar degrees of calculation volume too complex, leading to consumption of time too long, so in formed of each cluster in the select one and other text average similar degrees highest of text as a center text, after that the non-classified text calculate the similar degrees with various of center text, according to the calculation results to determine the non-classified text belongs to which cluster. If the distance between the non-classified text and all center text is larger than threshold T, the non-classified text will form a new cluster. On the contrary, the non-classified text will be merged into the minimum distance cluster, and the cluster will recalculate the Centre text.

A cluster formed on the second layer bolt node, and the cluster will be sent to the third layer bolt through Globlegrouping. When sending the cluster of second layer to the third layer bolt, the bolt node will conduct the Single-Pass cluster for the second time. Topology diagram as shown in Figure 2:

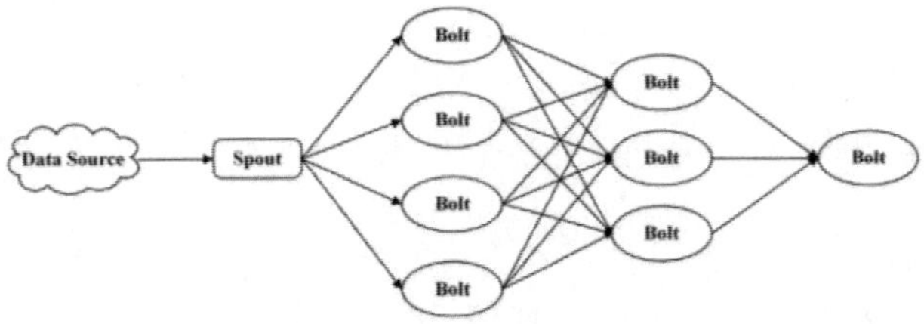

Figure 2. the topology of storm

3.3Pseudocode

```
BEGIN
    readData(input)        // Read data from the Hbase
    sendToFirstBolt(input)  // Sent from the spout to the first bolt node
    svm := nlpData(input)   // Data to quantify
    sendToSecondBolt(svm) // Quantitative data will be sent to the second bolt node
    C := nearest miroCluster  // Find the nearest micro-clusters
    maxSimilarityValue := similarityValue(p,C)
    if maxSimilarityValue > threshold then
            insertMiroCluster(p) // Similar to that of the most micro-clusters
    else
            create new miroCluster C// New micro-clusters
    end if
    sendToThirdBolt(C)        // The formation of micro-clusters is sent on to the   third
                                      floor bolt node
    T := nearest topic
    maxSimilarityValue := similarityValue(C,T)
    if C is new then          // Micro-cluster is first formed
            if maxSimilarityValue > threshold then
                    insertTopic(C)
            else
                    create new Topic T
            end if
    else
            if maxSimilarityValue > threshold then
                    if oldTopic(T) then     // If is the subject of one of the old topics
                            updateTopic(T)
                    else
                            splitOldTopic(T)
                            insertNewTopic(T)
                    end if
            else
                    splitOldTopic(T)
                    create new Topic T
            end if
    end if
END
```

4.Experiments

Testing the accuracy of text clustering, the paper uses NIST establishing evaluation criteria for TDT [10,11], a comprehensive performance evaluation, the formula is as follows:

$$C_{Det} = C_{Miss} \times P_{Miss} \times P_{target} + C_{FA} \times P_{FA} \times P_{non\text{-}target} \quad (4)$$

C_{Miss} as the coefficient of miss rate, C_{FA} as the coefficient of mistake rate, TDT evaluation values were 1.0 and 0.1 respectively; P_{Miss} as the miss rate; P_{FA} as the mistake rate; P_{target} and $P_{non\text{-}target}$ as prior objective probabilities, P_{target} =0.02,Pnon-target = 0.98. In order to further evaluate the TDT system performance, it is often used the C_{Det} generalized, the normalized error recognition price $(C_{Det})_{Norm}$, defined as follows:

$$(C_{Det})_{Norm} = \frac{C_{Det}}{\min\left(C_{Miss} \times P_{target}, C_{FA} \times P_{non\text{-}target}\right)} \quad (5)$$

The data set used in public opinion by the Guizhou Research Center, it contains 40 million data. This paper selects 10,000 test data from the date set. The selected data is related to hundreds of events. Text clustering similar threshold T is set to 0.05.

In order to test the speed and precision of S-Single-Pass algorithm, the paper adopted conventional stand-alone Single-Pass cluster and S-Single-Pass algorithm. In the S-Single-Pass algorithm ,the cluster contains a total of 16 nodes, including 1 spout node in the first layer of cluster, and the number of 1、3、5、7 nodes in the second and third layer respectively, and 1 bolt node in the final layer. The threshold T_c of the number of clusters is set to 2000, the threshold T_t of the number of topic is set to 3000. When the number of topic is larger than the threshold T_t, the cluster will remove the topic which is the least recently used. The value of Spout send queue is set to 10000. The result of calculating and the velocity contrast diagram shown in Figure 3:

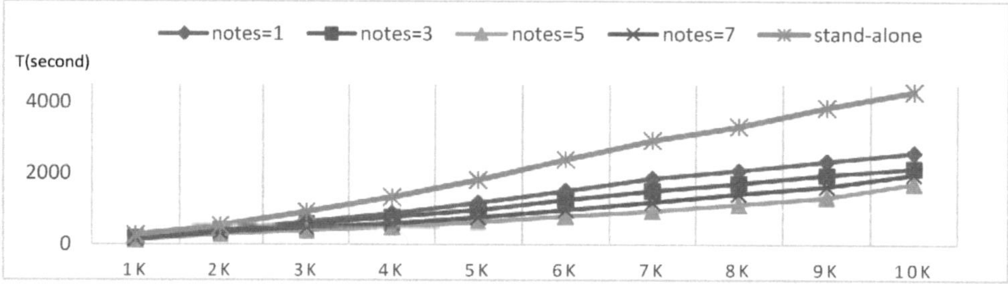

Figure 3. Speed comparison chart about two algorithm

In figure 3: When the number of nodes is 1, and the average of running speed is 1.66 times as much as the speed of stand-alone. On the whole, the efficiency increased by 66%.With the increase of the number of nodes in the second layer and the third layer, the program runs faster and faster. When the number of node in the second layer and third layer is up to 5 respectively, the total number of started node is up to 12, the program runs fastest, the efficiency increased by 151% overall. When the number of node in the first layer and second layer is up to 7 respectively, the total number of started node is up to 16, that is to say, all of the nodes have been started, the program runs slower and throughput is reduced. This is because when starting all of nodes, computer overload, information transfer between servers slow, then leading to the speed reduced.

Afterwards, compared to the accuracy of S-Single-Pass algorithm clustering and the accuracy of Single-Pass algorithm clustering, the paper conducted simulation operation. The algorithm different nodes mistake rate, miss rate, normalization error recognition costs as shown in table 1:

Table 1. Accuracy of contrast about two algorithm

Nodes	P_{Miss}	P_{FA}	$(C_{Det})_{Norm}$
1	0.2491	0.0457	0.4730
3	0.2519	0.0491	0.4925
5	0.2489	0.0401	0.4454
7	0.2569	0.0414	0.4598
stand-alone	0.2464	0.0386	0.4358

As shown in table 1, the different node number of the S-Single-Pass algorithm and Single-Pass make a comparison through P_{Miss}, P_{FA}, $(C_{Det})_{Norm}$ of three parts, it can be seen that different node number of the S-Single-Pass algorithm precision floating up and down is not large, basic flat, which can obtained based on Storm distributed framework of Single-Pass algorithm in guarantee precision of while, speed up the operation speed of algorithm.

5.Conclusion

Based on the poor efficiency of traditional Single-Pass algorithm employed in the massive data, this paper proposed a clustering algorithm based on distributed Single-Pass Storm to improve topic detection and tracking of timeliness. The structure of algorithm is use bilayer clustering, ensuring the accuracy of the clustering at the same time, and improving the speed of the cluster. Test results further validate the Single-Pass algorithm based on distributed framework for Strom to complete large-scale data processing in real time.

Acknowledgments

I would like to express my gratitude to the Beijing University of Chemical Technology for the financial support of this research. This study was supported by the Fundamental Research Funds for the Central Universities (JD1608) and the Fundamental Research Funds for the Central Universities (JD1502).

1. References

[1] Zhu, H., University, N., & Nanjing. (2011). Study on web topic online clustering approach based on Single -Pass algorithm. *New Technology of Library & Information Service*.

[2] Zhao, F., Lin, S., Gao, X., & Computers, S. O. (2016). The research and application of storm framework for large data. *Microcomputer & Its Applications*.

[3] Huang, S., Yang, Y., Li, H., & Sun, G. (2014, December). Topic detection from microblog based on text clustering and topic model analysis. In*Services Computing Conference (APSCC), 2014 Asia-Pacific* (pp. 88-92). IEEE.

[4] Cheng, X. Q., Jin, X. L., Wang, Y. Z., Guo, J. F., Zhang, T. Y., & Li, G. J. (2014). Survey on big data system and analytic technology. *Journal of Software*.

[5] Dong, X., Li, R., Zhou, W., Wang, C., Xue, Z., & Liao, D. (2013). Performance optimization and feature

enhancements of hadoop system.*Journal of Computer Research & Development*.

[6] Ke, M. A., & Lingjuan, L. I. (2016). Distributed real time stream data clustering algorithm and its implementation based on storm. *Journal of Nanjing University of Posts & Telecommunications*.

[7] WANG, M., YUAN, S., ZHU, Y., & WANG, D. (2014). Real-time clustering for massive data using Storm. *Journal of Computer Applications*, *11*, 006.

[8] Li, J. J., Cui, J., Wang, D., Yan, L., & Huang, Y. S. (2011). Survey of mapreduce parallel programming model. *Tien Tzu Hsueh Pao/acta Electronica Sinica, 39*(11), 2635-2642.

[9] Guo, Q. (2008). The similarity computing of documents based on vsm. ,*5186*(11), 585-586.

[10] NIST.The 2002 Topic Detection and Tracking (TDT2002) Task Definition and Evaluation Plan [EB/OL]. (2002)[2008]. ftp: //jaguar. ncsl. nist . gov //tdt/td t2002/evalplans/ TDT02 . Eval. Plan. v1. 1. ps .

[11] Seo, Y. W., & Sycara, K. (2004). Text clustering for topic detection.

IOP Conf. Series: Journal of Physics: Conf. Series **806** (2017) 012018 doi:10.1088/1742-6596/806/1/012018

An Inverse Kinematics Solution Based on CA-CMAC and ILC for Trajectory Tracking of Redundant DOF Manipulator

LIU Hua-qing[1]and LV Wang[1,2]

(1. Shanghai Institute of Satellite Engineering, Shanghai 200240, China; 2. School of Aerospace, Tsinghua University, Beijing 100084, China)

liuhuaqing@zju.edu.cn ; lvwang509@126.com

Abstract. The composite control strategy of CA-CMAC and ILC is adopted to solve the inverse kinematic problem of the redundant DOF manipulator during its real-time and high-precision tracking on the three-dimensional space target trajectory. A direct inverse model control strategy is adopted, in which CA-CMAC takes the current joint angles and the desired position increment of the manipulator as the input, and estimates the expected joint angle increments of the manipulator using the system history control experience. Then the estimated joint angles are taken as the initial value of the ILC module by which the control effect is improved iteratively. Based on the MATLAB, the tracking controls of linear and circular space target trajectories were simulated respectively. The results show that CA-CAMC and ILC composite control has better tracking precision and stability than CMAC control, while keeping the joint angles of the manipulator continuous and smooth during trajectory tracking.

1. Introduction

Inverse kinematics problem is a kind of important problem of the motion control of manipulators. Especially for the redundant DOF manipulator, the inverse kinematics problem often does not have the unique solution. At present, the method of solving the inverse kinematics problem of the manipulator can be classified into the following categories: analytic method [1], geometric method [2], iterative method [3], intelligent algorithm [4,5]. Although the analytic method can give all the roots, but the solution process is cumbersome and computationally large. The geometry method is simple, but it is only valid for some simple arms. Iteration method is suitable for most cases, but cannot provide all solutions, the calculation is huge and there is the possibility of non-convergence. Intelligent algorithms such as genetic algorithms, artificial neural networks, etc., due to real-time, stability and accuracy could not be guaranteed, so is rarely used.

The Cerebellar Model Articulation Controller (CMAC) proposed by Albus JS [6] in 1975 is a neural network model to simulate cerebellar function based on the Eccles cerebellar space-time model. CMAC is a tabular search-based adaptive neural network which can express complex nonlinear functions. It has the inherent capability of local generalization, information classification and storage, and has the advantages of fast learning speed and simple network structure. Because of the superior performance mentioned above, CMAC has better non-linear approximation ability than general neural network. On this basis, Sun [7] proposed the CA-CMAC (Credit assigned CMAC). Through the effective use of historical learning information, the learning speed is further improved.

Iterative Learning Control (ILC) was proposed by the Japanese scholar Arimoto [8] in 1984 for the repetitive motion of manipulator. The basic idea of ILC is to track the trajectory of the manipulator repeatedly. Based on the previous experience of the control system, the control variables can be modified according to the error between the actual trajectory and the expected trajectory. And gradually find an ideal input curve, making the actual trajectory of the robot arm close to the desired trajectory. ILC is applicable to the controlled object with repetitive motion properties. Theoretically, it can achieve complete tracking, and the computation is small, so it is suitable for the control of fast motion. However, ILC is sensitive to the initial state. Theoretical studies generally assume that the state of the manipulator is exactly equal to the desired state at the initial moment, or that the error is small. However, in practical applications, the initial error may be relatively large, in this case how to improve the ILC so that it can converge to the desired value, is an important issue. Many scholars have studied the ILC initial value problem, and gives the solution. Arimoto [9] studied the effect of initial positioning error on system tracking performance. Porter B [10] introduced the pulse correction to eliminate the initial error, thus achieving a complete tracing of the desired trajectory. Sun [11] proposed a learning algorithm with initial correction and terminal attraction to achieve complete tracking of the desired trajectory on the predetermined interval in the presence of initial error.

Aiming at the real-time, stable and high-precision requirement of inverse kinematics solution in the trajectory tracking control of the redundant manipulator, considering the CMAC control precision is not high enough but it has the advantage of real-time and generalization ability, while the ILC control precision is high but there is an initial value problem, a CA-CMAC and ILC composite controller is designed to realize the real-time and high-precision inverse kinematics tracking control of the redundant DOF manipulator to the target trajectory by complementing each other.

2. Control object description

The research object of this paper is redundant DOF manipulator such as the industrial manipulator with six degrees of freedom as shown in Fig. 1 (a). To simplify the analysis while retaining the ability of the manipulator to track target trajectory in three-dimensional space, two degrees of freedom of the wrist are neglected, thus, it is abstracted as the four-joint serial manipulator structure shown in Fig. 1 (b). Regardless of the posture constraints of the end-effector, the manipulator has four degrees of freedom and the workspace dimension is three, so it is a redundant manipulator.

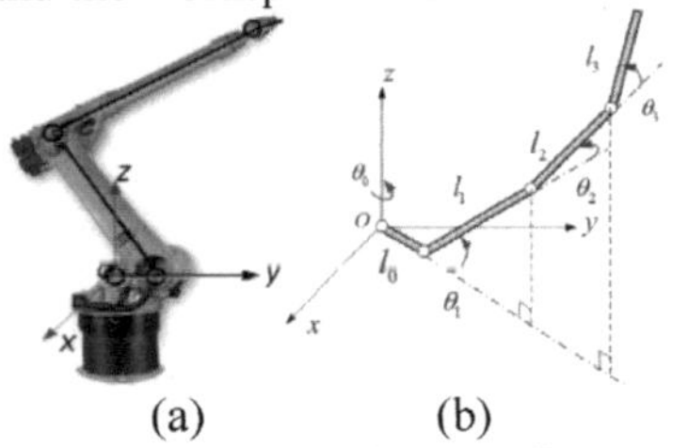

Figure 1. Sketch of the manipulator. (a): a common industrial manipulator; (b): the object of this study and its definition of degrees of freedom and other parameters

As shown in Fig. 1 (b), the coordinate system $o-xyz$ is fixed to the base of the manipulator, and the arm lengths of the four joints are l_0, l_1, l_2, l_3 respectively. The joint l_0 can be rotated around the z-axis, its joint angle is denoted as θ_0, and the initial position is directed to the x-axis positive direction. The joint l_1, l_2, l_3 respectively, has and only has one degree of freedom, and the three joints always move in the same plane, the corresponding joint angles are denoted as $[\theta_1, \theta_2, \theta_3]$, and the initial position of each points to the extension line direction of the previous joint. Thus, we could obtain the position of end-effector the manipulator:

$$\mathbf{R}=\begin{bmatrix} X \\ Y \\ Z \end{bmatrix}=\begin{bmatrix} [l_0+l_1\cos\theta_1+l_2\cos(\theta_1+\theta_2)+l_3\cos(\theta_1+\theta_2+\theta_3)]\cos\theta_0 \\ [l_0+l_1\cos\theta_1+l_2\cos(\theta_1+\theta_2)+l_3\cos(\theta_1+\theta_2+\theta_3)]\sin\theta_0 \\ [l_1\sin\theta_1+l_2\sin(\theta_1+\theta_2)+l_3\sin(\theta_1+\theta_2+\theta_3)] \end{bmatrix} \tag{1}$$

The goal of the trajectory tracking kinematic control of the manipulator is to control the joint angles $[\theta_0, \theta_1, \theta_2, \theta_3]$ so that the position of the end-effector $\mathbf{R}_d$ tracks the target trajectory in three-dimensional space in real time and accurately. Since the four joints are always in the same plane, the joint angle θ_0 has a unique solution:

$$[X_d \quad Y_d] = [\cos\theta_0 \quad \sin\theta_0]\sqrt{X_d^2 + Y_d^2} \tag{2}$$

Thus, the original problem is converted to the inverse kinematics problem of the planar three-joint manipulator, where the coordinate of the target trajectory is changed to $\mathbf{r}_d$:

$$\mathbf{r}_d = [x_d \quad y_d]^T = \left[\sqrt{X_d^2 + Y_d^2} - l_0 \quad Z_d\right]^T \tag{3}$$

The control variable is $\theta_1, \theta_2, \theta_3$. The actual position of the end-effector is:

$$\mathbf{r} = \begin{bmatrix} x \\ y \end{bmatrix} = \begin{bmatrix} l_1\cos\theta_1 + l_2\cos(\theta_1+\theta_2) + l_3\cos(\theta_1+\theta_2+\theta_3) \\ l_1\sin\theta_1 + l_2\sin(\theta_1+\theta_2) + l_3\sin(\theta_1+\theta_2+\theta_3) \end{bmatrix} \tag{4}$$

The control problem can then be described as follows: by controlling the three joint angles so that the actual position of the end-effector of the manipulator tracks the target trajectory, i.e. the tracking error tends to zero.

3. Design of the CA-CMAC and ILC Composite Controller

The composite controller consists of a CA-CMAC module and an ILC module, as shown below.

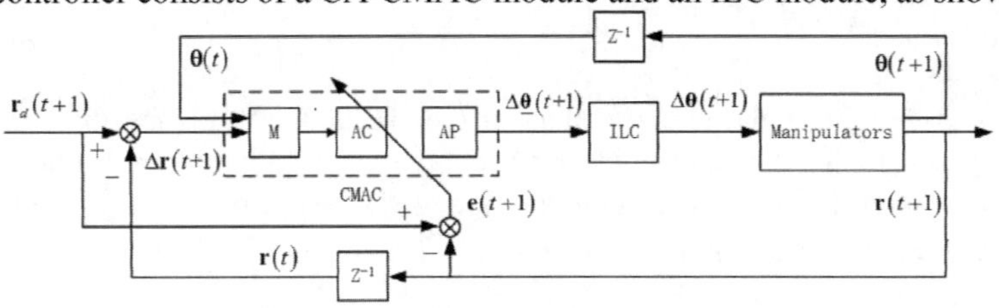

Figure 2. Control system block diagram

The control system works as follows: At control period t, the expected position of the end-effector of the manipulator $\mathbf{r}_d(t+1)$ is taken as the input signal of the system. $\mathbf{r}_d(t+1)$ minus the actual position of the end-effector, and the desired position increment $\Delta\mathbf{r}_{t+1} = [\Delta x_{t+1}, \Delta y_{t+1}]$ is obtained. The actual joint angles $\boldsymbol{\theta}(t) = [\theta_{1t}, \theta_{2t}, \theta_{3t}]$ of the manipulator at period t and $\Delta\mathbf{r}_{t+1} = [\Delta x_{t+1}, \Delta y_{t+1}]$ are merged into a vector C as the input to the CA-CMAC neural network controller (p is the proportional coefficient):

$$\boldsymbol{S} = [\theta_{1t}, \theta_{2t}, \theta_{3t}, p\Delta x_{t+1}, p\Delta y_{t+1}] \tag{5}$$

3.1. CA-CMAC controller

CA-CMAC is a kind of look-up table adaptive neural network which uses linear structure to express complex nonlinear function. It stores the learning data in the memory and outputs the sum of the weights of the activated units. CA-CMAC network consists of input space S , quantization space M , virtual memory space AC, the actual storage space AP and output layer Y . The mapping between the four layers is as follows [6].

3.1.1. S to M mapping. The continuous variable s is discretized as:

$$m_k = \left\lceil \frac{s_k - s_{k,\min}}{s_{k,\max} - s_{k,\min}} Q_i \right\rceil, k = 1,2,...,5 \tag{6}$$

where Q_k is the quantization step number; s_k is the kth element of the input vector; and $\lceil \ \rceil$ is the round-up operator.

3.1.2. M to AC mapping. The addresses of the activated units in the AC are coded as follows:

$$ac_i = \sum_{k=1}^{5} a_i^k + i \cdot Q^5, \quad i = 1,...,C \tag{7}$$

among them:

$$a_i^k = \left(\left\lceil \frac{m_i}{\Delta_k} + ofs(i) \right\rceil \backslash Q \right) Q^{i-1} \tag{8}$$

$$ofs(i) = (i-1) \cdot \frac{\Delta_k}{C}, \quad i = 1,...,C \tag{9}$$

The meaning of each symbol in the above formula is:
Δ_k : represents the quantization width of the k-th dimension input component s_k ;
$\backslash$: indicates the remainder operator;
Q : quantification level.

3.1.3. AC to AP mapping. Since the virtual storage space AC requires a large number of storage units, which is physically unrealistic, the storage addresses are compressed by the mapping. Using the hash coding technology to encode the unit address within the AP space ($ap_i, i = 1,...,C$ represents the address of the C activated C cells in AP; h is the hash coefficient):

$$ap_i = \left\lceil \left(\ln\left((ac+1) \cdot h\right) \backslash N_{AP} \right) \right\rceil, i = 1,...,C \tag{10}$$

Since the CA-CMAC output is the angular increment of the three joints, the weight matrix is designed as a three-column matrix. After the space compression, the actual storage space has only $3N_{AP}$ storage units.

3.1.4. AP to Y mapping. According to the unit address, read the network weights in the activated unit of the AP space, add and get the output:

$$\Delta\underline{\theta}_j = \sum_{i=1}^{C} w(ap_i, j) \quad , j = 1,2,3 \tag{11}$$

Where $w(ap_i, j)$ denotes the ap_i -th row and the j -th column of the weight matrix w .

3.2. ILC controller

To further improve trajectory tracking accuracy, an ILC controller is cascaded behind the CA-CMAC controller. At each control period t, the output value $\Delta\underline{\boldsymbol{\theta}}(t+1)$ is obtained by a series of mappings within the CA-CMAC controller. The value is then entered into the ILC controller as its initial value. The role of the ILC controller is to optimize iteratively around the initial value. The optimization objective function is defined as follows:

$$\min\ V\left(\Delta\boldsymbol{\theta}(t+1)\right)=\mathbf{e}^T\mathbf{e}/2 \tag{12}$$

In this optimal target sense, the optimal joint angle increments are the set of values that minimize the cost function, denoted as $\Delta\boldsymbol{\theta}=\left[\Delta\theta_1,\Delta\theta_2,\Delta\theta_3\right]^T$. The steepest gradient descent method is used to seek the optimal joint angle increments by a number of iterations:

$$\Delta\boldsymbol{\theta}^{k+1}(t+1)=\Delta\boldsymbol{\theta}^k(t+1)-\eta'\left(\frac{\partial V}{\partial\Delta\boldsymbol{\theta}^k(t+1)}\right)^T=\Delta\boldsymbol{\theta}^k(t+1)+\eta'\left[\mathbf{e}^k(t+1)\right]^T\mathbf{J} \tag{13}$$

It can be seen that the above equation is a P-type learning control law. Where $k=0,1,2...$ is the number of iterations; η' is the learning rate of ILC, taken as a positive value, to prevent divergence, the learning rate should be less than one; $\mathbf{J}$ is the Jacobian matrix defined as $\mathbf{J}=\partial\mathbf{r}/\partial\boldsymbol{\theta}$.Theoretically, after a sufficient iteration, $\Delta\boldsymbol{\theta}^k$ will completely approach the ideal value , but in view of the real-time control, only a limited number of iterations are implemented.

3.3. CA-CMAC weight adjustment algorithm

At the end of each control period, the weights of the CA-CMAC network should be adjusted. The purpose of weight adjustment is to improve the control precision, and finally to make the optimization objective function based on the trajectory tracking error converge to the minimum. The manipulator end-effector position corresponding to the joint angle increment $\Delta\boldsymbol{\theta}$ is $\mathbf{r}(t+1)$, and the error of $\mathbf{r}(t+1)$ with respect to the target track position $\mathbf{r}_d(t+1)$ is $\mathbf{e}(t+1)$. The CA-CMAC network weight is adjusted according to the tracking error. The adjustment algorithm is as follows: firstly, we use the sum of squares of errors to define the index function $V=\hat{\mathbf{e}}^T\hat{\mathbf{e}}(t+1)/2$, thus the deviation of the index function to the network weights is:

$$\frac{\partial V}{\partial w}=\hat{\mathbf{e}}^T\frac{\partial\hat{\mathbf{e}}(t+1)}{\partial w}=\hat{\mathbf{e}}^T\frac{\partial\hat{\mathbf{e}}(t+1)}{\partial\hat{\boldsymbol{\theta}}}\frac{\partial\hat{\boldsymbol{\theta}}(t+1)}{\partial w}=-\hat{\mathbf{e}}^T\mathbf{J} \tag{14}$$

The conventional weight adjustment algorithm distributes the correction values evenly to the activated physical storage cells so that the activated cells that do not need to adjust the weights excessively are adjusted repeatedly due to the principle of average distribution, while those that require multiple adjustments has not been fully adjusted, resulting in low learning efficiency. In order to improve the learning efficiency, the CA-CMAC was proposed [7]. The historical information of the physical storage unit determines the magnitude of its weight adjustment: for any unit, the more it is activated in the past, the higher its credibility and should be assigned a small correction value; on the contrary, the less it is activated in the past, the less reliable it is, and should be assigned a larger correction value. In this paper, the steepest descent method based on credibility assignment is used to adjust the weights, where the learning rate η is positive [7]:

$$w_{ij,t+1}=w_{i,t}-\eta\frac{C\left[f(i)+1\right]^{-1}}{\sum_{j=I}^{C}\left[f(j)+1\right]^{-1}}\frac{\partial V}{\partial w}=w_{i,t}+\eta\frac{C\left[f(i)+1\right]^{-1}}{\sum_{j=I}^{C}\left[f(j)+1\right]^{-1}}\mathbf{e}(t+1)^T\mathbf{J}\ ,i=1,...,C \tag{15}$$

Among then, I represents the set of C physical memory cells being activated and $f(i)$ represents the number of history activations of the ith activated physical memory cell.

4. Simulation verification

The arm lengths of the four joints are taken as $l_0 = 0.1\text{m}$, $l_1 = 0.5\text{m}$, $l_2 = 0.5\text{m}$, $l_3 = 0.1\text{m}$, respectively. The generalization factor of the CA-CMAC $C = 10$, the total number of cells in memory $N_{AP} = 3000$, the hash coefficient $h = 12345$. The quantization levels of each component of input variable is taken as 360, and the learning rate η=0. 1, and the learning rate of the ILC η'=0.5. Base on MATLAB platform, the tracking of linear and circular trajectories of the redundant DOF manipulator under the composite control of CA-CMAC and ILC is simulated. The control precision of a complete trajectory tracking control process is defined as the root mean square of the tracking error $\mathbf{e}$, noted as $\overline{\mathbf{e}}$. Let the four joint angles be zero at the initial moment, and after 1000 training periods, we get the simulation results as shown below.

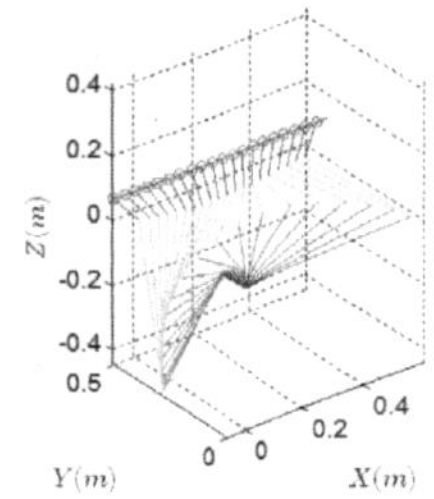

Figure 3. Linear trajectory tracking

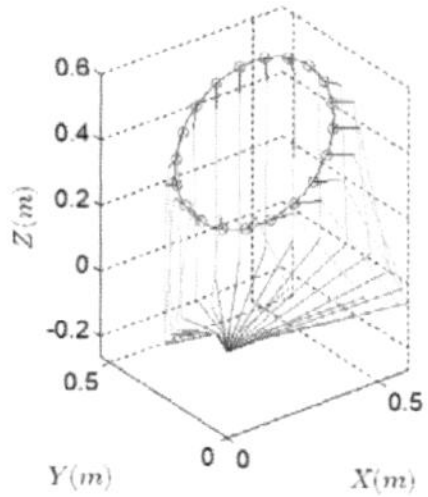

Figure 4. Circular trajectory tracking

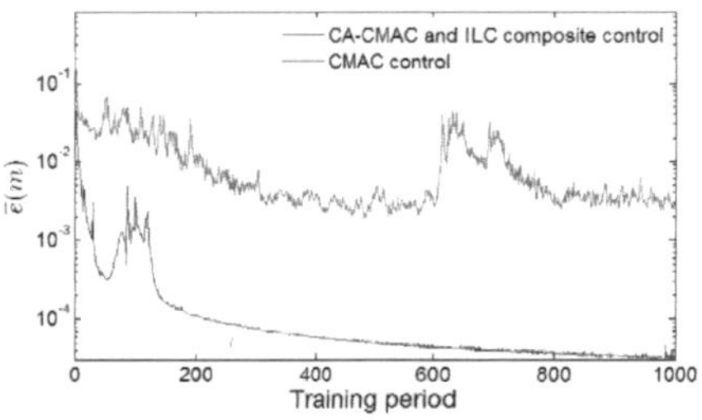

Figure 5. The optimization process of the linear trajectory tracking accuracy

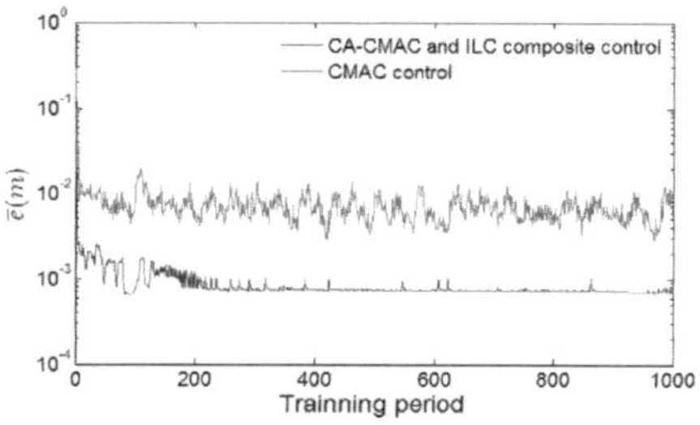

Figure 6. The optimization process of the circular trajectory tracking accuracy

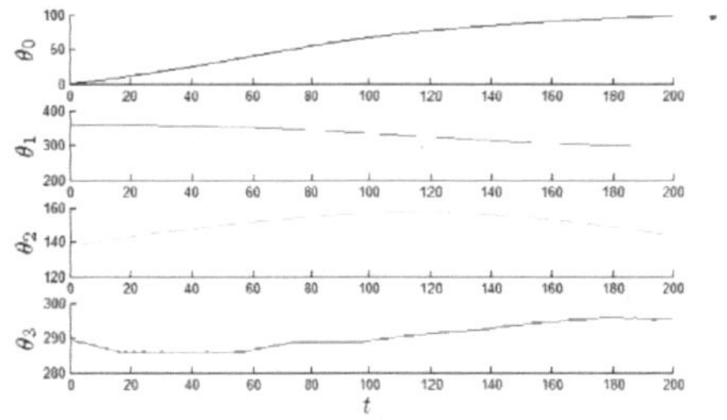

Figure 7. After training, the time-history curves of each joint angle in the tracking process of the linear trajectory (angle unit: °; time unit: dimensionless)

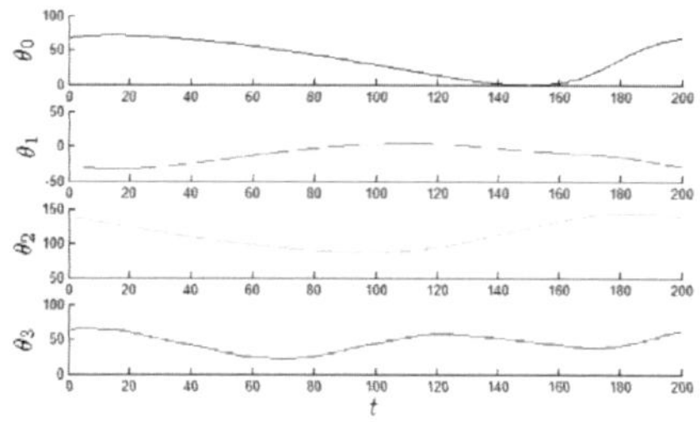

Figure 8. After training, the time-history curves of each joint angle in the tracking process of the circular trajectory (angle unit: °; time unit: dimensionless)

It can be seen from the Figs.5 and 6 that with the increase of the number of training periods, the tracking error of the manipulator to the target trajectory can be reduced gradually and stabilized to a certain level, which indicates that the composite controller has a certain convergence. Comparing the red and blue curves shown in Figs. 5 and 6, it can be seen that the accuracy of the composite control is more than one order of magnitude higher than that of the CMAC control, and the stability is also stronger. This is because: 1, the weight assignment algorithm based on the Credibility Assignment is adopted; 2, the addition of the ILC controller not only enhances the control algorithm stability, but also improves the trajectory tracking accuracy. It can be seen from Fig.7 and Fig.8 that the time-history curves of the four joint angles in the two examples are all smooth, which indicates that the manipulator moves smoothly during trajectory tracking and the energy consumption is small. Once the weight adjustment is complete, when one input the target trajectory, the controller only needs to look up the table to get the appropriate control input of the manipulator, so it has good real-time performance.

5. Conclusion

The output of the CA-CMAC provides an initial iteration value for the ILC, which solves the problem of the ILC initial value. In addition, the cascade of the ILC to the CA-CMAC not only restrains the divergence phenomenon caused by over-learning of the CA-CMAC, enhances the stability of control, but also improves the control precision and realizes the complementary effect of the CA-CMAC and the ILC. Considering that the composite controller is simple and easy to implement, it has certain application value.

References

[1] Jiang HC, Liu SR, Zhang BT. Inverse kinematics analysis for 6 degree-of-freedom modular manipulator. Journal of Zhejiang University. Engineering Science. 2010 Jul;44(7):1348-54.

[2] Chen P, Liu L, Yu F, Li H, Yang F, Wang X. A geometrical method for inverse kinematics of a kind of humanoid manipulator. Jiqiren(Robot). 2012 Mar;34(2):211-6.

[3] Chen IM, Yang G, Kang IG. Numerical inverse kinematics for modular reconfigurable robots. Journal of Robotic Systems. 1999 Apr 1;16(4):213-25.

[4] Guez A, Ahmad Z. Solution to the inverse kinematics problem in robotics by neural networks. InNeural Networks, 1988., IEEE International Conference on 1988 Jul 24 (pp. 617-624). IEEE.

[5] Köker R, Öz C, Çakar T, Ekiz H. A study of neural network based inverse kinematics solution for a three-joint robot. Robotics and Autonomous Systems. 2004 Dec 31;49(3):227-34.

[6] Albus JS. A new approach to manipulator control: The cerebellar model articulation controller (CMAC). Journal of dynamic systems, measurement and control. 1975 Sep 1;97(3):220-7.

[7] Su SF, Tao T, Hung TH. Credit assigned CMAC and its application to online learning robust controllers. IEEE Transactions on Systems, Man, and Cybernetics, Part B (Cybernetics). 2003 Apr;33(2):202-13.

[8] Arimoto S, Kawamura S, Miyazaki F. Bettering operation of robots by learning. Journal of Robotic systems. 1984 Jun 1;1(2):123-40.

[9] Arimoto S. Learning control theory for robotic motion. International Journal of Adaptive Control and Signal Processing. 1990 Nov 1;4(6):543-64.

[10] Porter B, Mohamed SS. Iterative learning control of partially irregular multivariable plants with initial impulsive action. International Journal of Systems Science. 1991 Mar 1;22(3):447-54.

[11] SUN, Ming-Xuan, et al. "Feedback-aided PD-type Iterative Learning Control: Initial Condition Problem and Rectifying Strategies." *Acta Automatica Sinica* 1 (2015): 017.

Institute of Physics
Dirac House, Temple Back
Bristol BS1 6BE UK

ISSN: 1742-6588
ISBN 978-1-5108-3681-5

9 781510 836815